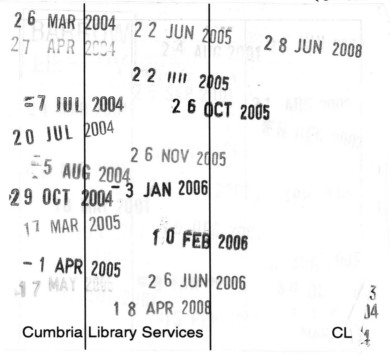

CANAL MANIA

CANAL MANIA

Anthony Burton

✣ 200 YEARS OF BRITAIN'S WATERWAYS ✣

with
Photographs by Michael Taylor
Drawings by Peter White

AURUM PRESS

FACING TITLE PAGE: *Bratch toll office, Staffs & Worcester.*

OVERLEAF: *Canal bridge, Kennet & Avon.*

CONTENTS

Preface

This book was suggested to me by British Waterways, as part of their Canals 200 celebrations, commemorating the bicentenary of the peak mania year of 1793. I should like to express my thanks to all those who helped in its production: to British Waterways' chairman, David Ingman; to Vanessa Wiggins, Roy Jamieson and Simon Salem; and to David McDougall of the National Waterways Museum at Gloucester. It was a particular pleasure for me to be able to collaborate on a book with Peter White, British Waterway's Chief Architect for over 20 years. We first talked about working together in this way in the early 1970s; we have finally made it in the 1990s.

Although British Waterways has co-operated in the production of this book, the text and the views expressed in it are my own: they are purely personal and do not represent any sort of official view whatsoever.

The illustrations for the book come from three very different sources. The historic black and white illustrations are from British Waterways Archive, the National Waterways Museum and David McDougall's personal collection. Peter White travelled to all the 1793 canals to produce his remarkable 'overview' drawings. The colour photography is rather different. Michael Taylor had no previous experience of waterways photography, and his only brief was to look at the canal system and respond to it whatever way he liked. The result is a series of fresh vivid and often dramatic pictures which should be thought of not as illustrations of the text but as a separate photo essay. I have looked at the canals with a new eye since seeing his work and Peter White's extraordinary bird's-eye views. I hope others will be similarly excited and will be encouraged not only to seek out some of the derelict, little-known canals described here, but also to return to old favourites with renewed enthusiasm.

ANTHONY BURTON
BRISTOL, 1992

THE

COMPANY OF PROPRIETORS

OF THE

Oakham Canal.

THESE are to Certify, T**HAT** *George Maule of Lincolns Inn London Esquire, the Nominee of his Majesty As Administrator of the Goods Chattels and credits of Henry Donley late of Oakham in the County of Rutland M.D. who died Intestate a Batchelor and a Bastard*

is on the Day of the Date hereof, a Proprietor of O**NE** S**HARE** of *One Hundred and Thirty Pounds* in the O**AKHAM** N**AVIGATION**, being

Number *177*, in the S**HARES** of the said Navigation, subject to the Rules, Regulations, and Orders of the said C**OMPANY**, and T**HAT** he, his Executors, Administrators, and Assigns are entitled to the Profits, and Advantages of such S**HARE**.

Given under the Common SEAL of the said Company the *Twenty second* Day of *May*, in the Year of our Lord, One Thousand Eight Hundred and *Twenty*.

Clerk to the said Company of Proprietors.

A MANIA FOR CANALS

Today the phrase 'a mania for canals' suggests an impassioned love of Britain's old waterway system. For canal maniacs – and I count myself among their number – there is a deep pleasure to be found in chugging at the steady speed of a none-too-brisk walker through quiet countryside, or in sidling up, almost unnoticed, among the back streets of town and city. It is an activity more than a little tinged with nostalgia for what can now seem to be a wonderfully unhurried way of life, free from the stresses of modern living. The fact that in reality it was a life of unremitting hard work for meagre financial reward does not diminish the warm glow of romance that surrounds the world of the working canals. They have become, to many, no more than quaint survivors from an earlier age. But it was not always so.

Two hundred years ago 'canal mania' had a very different meaning. Then it suggested a wild enthusiasm for a new and exciting mode of transport that was to carry the world to a Utopia, based on industry and trade. The canal maniacs of the 1790s had no reveries of gently gliding over placid waters; if they dreamed at all, their dreams were of boats weighed down with cargo and of balance sheets loaded with profits. No account of a boat gently sliding under high-arched bridges of mellow, red brick or of drifting on a summer's day under the cool branches of a softly leaning willow stirred their blood. Looking through the newspapers of the time, one soon finds examples of the sort of account that got their pulses racing.

The shares of the Birmingham Navigation originally cost £140. In December last, one of them sold for £1,080 – and last week £1,150 was offered for another share and refused.

Newspapers also described, with great enthusiasm, meetings to promote new canals: they were always 'numerous assemblies' and the gentlemen present inevitably 'respectable'. But behind the formula, one can still glimpse the real enthusiasm of

Canal shares were not cheap. This Oakham Canal share has a face value of £130; a speculator would hope to sell it for a great deal more.

those days. In 1792 the prospective investors in the Grand Junction Canal were invited to gather at an inn at Stony Stratford, but so many turned up that the meeting had to be moved to the parish church. The promoters were asking for £250,000, but one million pounds was promised there and then by the members of this unusual congregation. A fortnight later, those who met to invest in the Leicestershire & Northampton Union found that not even the church would hold them, and they had to make do with a field.

These were heady days in the history of canals, but the beginning was altogether more modest. In 1760 Parliament gave its approval for a canal, scarcely half a dozen miles long, that was to take coals into Manchester from the mines at Worsley. The boats were plain and unadorned; the cargo scarcely exotic. Yet the Bridgewater Canal gripped the popular imagination. Partly it was the novelty of a wholly artificial waterway that showed its disdain for its natural brethren by soaring haughtily right over the River Irwell on an aqueduct. The notion of boat passing over boat was new and exciting. Aristocratic tourists made detours from the Lake District or from the spas and hills of Derbyshire to view the new phenomenon. They penned limp-wristed accounts of the 'navigable canal in the air' across which, one commentator wrote, 'I durst hardly venture to walk, as I almost trembled to behold the large River Irwell underneath me.' Others took a more phlegmatic and practical approach to the canal, reserving their enthusiasm for the account books, which showed coal prices halved in Manchester and a steadily rising profit.

The Bridgewater Canal was an undoubted success, yet there was no great rush to follow this promising lead. The Bridgewater itself was extended, but six years were to go by before work began on any other British canals – and, in some cases, a great many more years were to pass before they were completed. The Trent & Mersey Canal was begun in high optimism in 1766 but was not ready until 1777; while promoters of the Leeds & Liverpool, which was started in 1770, had to wait until 1816 before they could celebrate the opening of the whole canal. In the 30 years following the Bridgewater, only 25 new projects were begun. Even allowing for the fact that these included long and important routes, it was scarcely a transport revolution. Then, in 1790, it all began to change. That year saw just one new name added to the list; but there were six in 1791, seven the year after, and in 1793 the movement peaked with no fewer than 21 new canals being authorized. The number fell again to 11 in 1794 and the decline continued. It was that one brief period, centred on 1793, that came to be known as the years of canal mania. But why did it take so long to arrive? Why did more than 30 years have to pass between the start of the Bridgewater and the full flowering of the canal age?

The Bridgewater Canal did not have an easy birth. In the early part of the century there had been an orgy of popular speculation, centred on the government-sup-

The Bratch, Staffs & Worcester.

ported South Sea Company. The whole precarious structure burst apart in 1720, bringing many investors – large and small – to ruin. The event became known to history as the South Sea Bubble. Bubbles, once broken, alas cannot be repaired, but Parliament tried to ensure that no such financial disasters would occur again. It enacted laws regulating joint stock companies, one result of which was that the Duke of Bridgewater, like all the promoters who followed him, had to have an Act of Parliament before he could build a canal. This was to prove a long and costly business, for he was bitterly opposed by the owners of the old river navigations, who were anxious to preserve a cosy monopoly. The Duke, a young man still in his twenties, financed the whole scheme himself and, as the Parliamentary expenses rose, saw his own funds diminished, so that by the time work could actually begin he was already sliding into debt. He was soon borrowing from the bank, from relations and friends. It was said that when the local parson wanted his debt repaid, the Duke went into hiding and was somewhat ignominiously run to ground in a hay loft.

The Bridgewater Canal was, in many ways, an unlikely precedent. Perhaps it needed the bold imagination of the pioneer, prepared to risk everything he owned in the venture, to establish the trail that others could follow. It was unusual, too, in that the direction of the whole enterprise fell to just three men: the Duke himself, his very astute agent, John Gilbert, and a semi-literate, middle-aged millwright, James Brindley. It is not surprising to find that the rest of the world was happy to sit back and see how the triumvirate would fare. Would the Duke recover his fortune? Would the new system devised by Gilbert and Brindley stand the test of time?

In the event, the Duke's investment more than proved its worth and the 'castle in the air', as sceptics had dubbed the aqueduct across the Irwell at Barton, proved wholly reliable. The success inspired men like Josiah Wedgwood, the great potter who had himself begun to build a substantial business from modest beginnings, to turn their attention to canals. He himself actively promoted the Trent & Mersey Canal, a huge leap in scale from the little Bridgewater. This canal, aptly known as the Grand Trunk, was to sweep across the face of England, uniting two great navigable rivers. It was to be over 93 miles long and was to feature a tunnel that was to pass for a mile and a half under Harecastle hill. Nothing of the sort had even been attempted before. This was a scheme of a daring to match the Bridgewater and its Barton aqueduct. Other early schemes were no less exciting. The great Scottish rivers, the Forth and Clyde, were to be joined by a broad canal, and the Leeds & Liverpool was to force a way through the Pennines. Then, just as the whole process of canal-building was gathering momentum, the brake was applied and everything all but came to a halt.

It was not a lack of confidence that created the problem, but a lack of cash – and it had nothing to do with the viability of the canals as such. The problem was not even born in Britain at all, but across the Atlantic, where the American colonists demanded independence and went to war to acquire it. A colonial struggle in a distant country might have had little effect, but then other European powers were drawn into the conflict and trade was devastated. Britain lost the war, but not the peace. The empire, with the exception of the new United States, remained not only intact but, if anything, more soundly based. And though colonies had been lost in America, a new trading partner had been formed, whose cotton fields were ultimately to feed Britain's growing textile industry. Once the dust of war had settled, the peace talks of 1782–3 were seen as by no means unfavourable. Fortunes were slowly rebuilt and, just as important, confidence began to grow again. Even the French Revolution of 1789 had no immediate effect – it gave Britain's leading trade rival on the world stage more than enough preoccupations at home. And in Britain, the Industrial Revolution was moving forward with ever-increasing speed: work was moving from cottage to mill and factory; power went from human hand to water-wheel to steam-engine; the population began to shift from country to town. One commodity was everywhere in demand, coal – coal to feed the steam-engines, to heat the homes of the new towns and cities, to be turned to gas to light the streets and works. And coal was the ideal commodity for loading on boats to be hauled along the canals.

The mania of the 1790s had its origin in a real need for an efficient transport system to meet the demands of a changing world. That, however, was only part of the story. Alongside the industrialists and merchants, who saw the canals as a necessary part of a developing industrial world, were the speculators, who saw them as a means of making a quick profit. At the meeting in July 1792 to start the promotion of the Grand Junction Canal, only those who were actually in the church at Stony Stratford could subscribe – no proxy votes were allowed. Once a subscriber had his name on the list, it entitled him to shares – and he could begin trading these promised shares virtually immediately. It was a situation not unlike privatization in the 1980s, where those who got in at the beginning could cash in their shares for an instant profit. In the case of canal shares, there was no need even to wait for the canal to be authorized – there were investors more than ready to grab at the chance to be in line for shares in any new scheme that came forward. Not surprisingly, the enthusiastic speculator found himself dashing all over the country in order to get his hands on that magical piece of paper that said he was entitled to shares. One commentator, looking back on those days, wrote of the many who were prepared to sleep in barns and stables in order to be ready for a subscription.

Companies, good, bad and indifferent, found it equally easy to raise cash for their schemes, and how were the enthusiasts to tell one plan from another? The list

Company seals were intended to express the grandeur of the concern: the Grand Junction above; the Leicester Navigation below.

of promoters was of little help. In 1792 a number of eminent men appeared in the announcement for a public meeting to promote what was to become the successful Gloucester & Berkeley Ship Canal. At the same time, the same names appeared on another announcement promoting a canal from Bristol to Taunton – destined never to be built at all. For every sound scheme that was to go through to completion, there were others where money would be spent on Parliamentary business, but which would fail at this very first obstacle. So much cash was involved that Parliament, remembering the disastrous South Sea Bubble, raised the matter. In March 1793 there was a debate over whether dealings in canal shares should be limited, and whether or not something should be done to curb the very high profits being made by canal companies as well as by share dealers. A number of members expressed alarm at the way canals seemed to be covering the entire country.

One Hon. Member wished his grand-children might be born web-footed, that they might be able to live on fish, for there would not be a bit of dry land in this island to walk on.

And a proposal was put forward to stop all work on canals in harvest time, as so many workers were leaving the land to take up work as navvies. In the end, Parliament decided to do nothing at all: the navvies were to go on digging, the speculators to continue speculating.

It is tempting to look at these brief years of hectic activity and see nothing but the greedy on the hunt for quick profits. In fact, they represented a minority of investors. Canals were still being promoted by local interests, which saw them as good, long-term projects that would help bring trade and prosperity to a region. Inevitably some promoters paid scant attention to the practicality of their schemes. Canals were planned, blithely ignoring hills and rivers that lay across their route; waterways were proposed with no thought of where the water supplies were to come from to fill them and keep them full. Accounts of the trade to be carried bore scant resemblance to the realities of trade in a region. These were the schemes that quickly faltered and died, and a few hapless investors saw their dreams of riches expire with them. The £100 share was not going to fetch £1,000 at auction – it had less value than yesterday's newspaper. Even those schemes that did go forward were likely to come up against a problem that has probably been a feature of engineering works since the pharaohs began ordering pyramids: estimates of costs fell far below the actual costs that had to be met. An Act of Parliament would specify the amount that could be raised, and if that was spent, then a new Act had to be acquired to raise more. The Gloucester & Berkeley set off in 1793 with permission to raise £200,000: five Acts and 32 years later the final £50,000 was authorized, mak-

*Lock-widening at Braunston
c.1930, as part of the Grand
Union improvement scheme.
Ways of working have scarcely
changed since the days of the
canal mania.*

ing a grand total of £480,000. Even so, schemes were put forward at the height of the mania that were to prove their worth over the years, and if the age of big, quick profits had gone, the age of good, steady returns became a reality for many.

The mania years were, in any case, never just about money. They were years that saw a new age of engineering come into being. Techniques developed during the period that not only lasted for the rest of the canal age, but were carried forward, with surprisingly few changes, into the railway age that followed. It was in these years that the framework that had been established in the first rush of enthusiasm was filled in to create a network that served the country well.

The main part of this book is based on the canals of that one crucial year at the height of the mania, 1793. Looking through the list of canals authorized then, it is extraordinary to see not just how much variety there is, but how every type of canal is covered, from the private waterway serving a single industry to the ship canal that would create a whole new inland port. The canals of that year have also had very different fates: some have become little more than a faint mark on the landscape, like a faded photograph in a long-abandoned album; others are bustling with boats, even if those boats now contain holiday-makers rather than coal or pig-iron; a few still have commercial traffic – and the potential to carry a great deal more. But before one can understand this second generation of canals, one must first look back to the generation that preceded them.

THE AGE OF BRINDLEY

James Brindley dominated the early years of canal-building, yet it is hard to imagine a less likely character to fill that role. He was born to a poor family in Derbyshire in 1716, and as soon as he was old enough he went out to work, with an education that was at best rudimentary. In 1733 he was apprenticed as a millwright to a master who turned out to be more addicted to the bottle than the mill, so that young Brindley had to pick up his skills as best he might. This seems to have had a profound effect on his personality. Forced at an early age to rely on his own efforts, he showed himself in later life notably unwilling to listen to the advice of others. It was said that when he had a particularly difficult problem to solve in his canal engineering days he would retire to a darkened room to think, and not emerge until he could sketch out his solution. He was a practical, bluff, unromantic man. He was married in 1762 to the young daughter of his chief surveyor, John Henshall and his bride, just out of school, was taken to a house where the view was of the men at work on the Harecastle tunnel. On one occasion, when Brindley was in London on Parliamentary business for canal promotion, he was persuaded to go to the theatre, but he found the emotional temperature far too high, and the whole affair upset him so much that he vowed never to go again – nor did he.

It is easy to characterize Brindley as a coarse, uneducated man who just happened to be blessed with an innate talent for laying out canals: but this would be to underrate the man. Early in his career he showed a natural inventiveness, trying his hand at all kinds of inventions from steam engines to silk-spinning machinery. The Duke of Bridgewater did, after all, have a free hand in the choice of engineer for his project, and the fact that he selected Brindley is itself a mark of his reputation and abilities. It is not the only sign. Josiah Wedgwood was a man who had built up his world-famous enterprise from comparatively modest beginnings – but he was also a man who mixed with many of the foremost intellectuals of the day, men such

Bow-hauling a Thames & Severn barge near Sapperton tunnel.

as the poet, philosopher and scientist Erasmus Darwin, and the radical scientist Joseph Priestley. That Brindley was considered by Wedgwood the equal of such august company is as clear a mark as any of his worth. Wedgwood had no doubts at all about Brindley's abilities – he was, quite simply, a 'genius'.

James Brindley is a man about whom one longs to know more. Some of his contemporaries never looked beyond his rough clothes and rougher speech, and characterized him as 'a boor'. Others saw behind the crude exterior and glimpsed a man of sensitivity and great originality. But if we lack the written documents on which traditional biographies are built, we do have an eloquent testimonial in his work. The Bridgewater Canal is the obvious starting place, yet somehow one never feels quite at ease with it. True, contemporaries heaped praise on Brindley's head as the sole author of this remarkable work. Nevertheless, it seems to have little in common with later works produced under Brindley's direction. Perhaps one should not be too surprised at this, if one looks at its origins and the personalities involved.

First, there is the Duke of Bridgewater himself. As a young man he did precisely what young aristocrats were supposed to do: he went off on the Grand Tour of Europe. He behaved impeccably, doing as others did, and buying up antiquities by the crateful, but his actual enthusiasm for the art of the ancient world can be judged from the fact that when those same crates reached England he never bothered to open them. He also did what his compatriots did not do: he went to see the industries of Europe. In particular he visited the Grand Canal of Languedoc, now known as the Canal du Midi. This canal, completed in 1681, was described by no less a person than Voltaire as *'le monument le plus glorieux par son utilité, par son grandeur, et par ses difficultés'*. Useful it certainly was; grandeur it had; and there was no shortage of difficulties to be overcome. It was 150 miles long and contained all the features that were later to be found in British canals – locks, aqueducts and a tunnel. The jeerings of the Duke of Bridgewater's insular critics must have seemed quite ludicrous to a man who had seen such a wonder and was himself preparing an altogether more modest endeavour. The Duke had seen not only what could be done, but what had been done almost a century earlier.

The next member of the triumvirate, John Gilbert, is a somewhat more shadowy figure, but he was a practical man of affairs, in charge of a considerable estate, which included the mines at Worsley. It was mining that brought Brindley to his notice, for the millwright had been involved in a considerable scheme, involving cutting extensive channels for water drainage at the aptly named Wet Earth Colliery. How much time Gilbert was able to spare for the problems of canal construction is not known, but it could not have been a great deal. His job as agent required his full-time attention before and after the canal-building period, so there is no reason to assume that it was any less demanding during those years. Nevertheless, he was a confident man, more than willing to support the venture in practical ways as well as providing moral support. On this canal, privately financed and privately run, the novice engineer was sure of his employer's backing for, and understanding of, everything that was done.

So to Brindley himself: no doubt he listened to the Duke's accounts of aqueducts and other marvels in France, but the information was of little help in deciding how such structures were to be built in practice. Even if accounts had been written, he would have been incapable of reading them. Across the channel, the major problems of canal construction had been solved, but for all the use they were, they might as well have been the canals of Mars. Brindley had to set about solving the problems all over again – but at least he knew that they could be solved. That was a great boost to his confidence. So he went to work devising techniques such as 'puddling' – mixing clay and water and spreading it along the bottom and sides of the canal to prevent it from leaking. The heavy clay lining had a direct effect on the construction of the famous Barton aqueduct. Not only had it to withstand the pressure of the water, but it also had to carry the weight of the 'puddle'. Add to that the

Barton aqueduct carrying the Bridgewater Canal across the River Irwell.

necessity for it to be carried on arches that would allow vessels to pass beneath it on the River Irwell, and it was clear that Brindley's, and Britain's, first aqueduct needed to be a massive and substantial structure. Viewed from the lower level of the Irwell, it was very similar to bridges of the period. The semi-circular centre arch had a span of 57 feet, with smaller arches to either side. It was a reassuringly solid rather than an elegant structure, and that solidity was even more apparent when the aqueduct was viewed from the level of the new canal. It was 36 feet wide, but only half of that width was waterway, the rest being taken up by puddle and packed earth to carry the towpath. Scornful critics of the Barton aqueduct were not to be given the satisfaction of seeing the 'castle in the air' crumble and fall. It was to stand until the end of the nineteenth century when it was demolished to make way for the Manchester Ship Canal, and Brindley's aqueduct was replaced by a swing aqueduct that could be moved to allow ships to pass through.

Brindley faced new problems when it came to extending the canal to the Mersey and Runcorn. The first canal had run lock-free from the mines at Worsley to the basin at Castleford. Now, however, Brindley would be faced with the necessity of building locks to unite canal and river. The Mersey was already busy with inland waterway craft, known as 'flats', generally 71 feet long and 14 feet 3 inches beam, but of varying draught. There were two choices – either to build locks that would take craft of this size, or to build on a more modest scale and be faced with the trouble and expense of exchanging cargo between river craft and canal boats. In the end, boldness won. The flight of ten locks was built to take the flats, including a new class of shallow draught that came to be known as 'Dukes'. But before this stage had been reached, Brindley had moved on to other schemes, under other directors, and was to take a quite different view of scales and dimensions.

In 1766 Acts were passed permitting the construction of two important canals – the Staffordshire & Worcestershire, and the Trent & Mersey – which would link together three of the great navigable rivers of England, the Trent, the Severn and the Mersey. There was also to be a connection with the Bridgewater Canal. The rivers of the time were busy with a variety of craft, but as each was self-contained there was no need to worry about whether the flats of the Mersey, for example, could use facilities designed for the trows of the Severn. The connecting links changed all that, and with plans already in existence to complete a crossway of canals that would also link to the Thames, it was clear that a decision taken for any one of these canals would affect the others. The obvious choice would seem to be to accept the dimensions of the Bridgewater and the Mersey as a standard. This would allow any vessel up to the size of a Mersey flat to travel throughout an extensive network of inland waterways. Why should such an obviously efficient plan not be put into practice? There were a number of arguments that could be raised in favour of less ambitious canals.

A broad canal not only requires broad structures – wider locks, bridges, and so on – but also a much wider strip of land than a narrow canal, all of which has to be paid for. Was it worth the effort? The eighteenth century was a great age for experimentation and for facts and figures. One experiment set out to show just how great a load could be hauled along a canal by a single horse. The answer came out at around 50 tons – or just about the load carried by the 'Dukes' on the Bridgewater Canal. Again this was an argument in favour of cost-cutting efficiency, getting the most out of boat and horse. But the same experiment also showed just how little could be managed by a horse on land – varying from a paltry one-eighth of a ton for a pack-horse to a scarcely dramatic two tons for a waggon on one of the new, well-surfaced turnpike roads. So even a comparatively small boat would still be a vast improvement over land carriage. There were other factors, however, beside loading figures that weighed on the engineer's mind. If wide boats were to be used, then everything on the whole system, without exception, had to be designed to accommodate them. A feature of the Bridgewater Canal was a labyrinth of underground workings that took boats deep into the heart of the mine system at Worsley. Here, the dark passages cut into the sandstone cliff were served by very basic, open double-ended craft. The ribs and keel were exposed to view, and this skeletal look earned them the name 'starvationers'. They were seven feet beam, half the width of the river barges. These would seem to have been the inspiration for Brindley's narrow-boat canals.

The connection between boats designed for use in the mines and those for use on a major inland waterway system may not be immediately obvious, but Brindley was now faced for the first time with the necessity of taking his new canals through tunnels. In particular, the Trent & Mersey was to pierce Harecastle hill north of Stoke in a 2,897-yard-long tunnel. Nothing of the sort had been attempted in Britain before – even on the Continent, nothing on quite the same scale had been tried. Had his locks allowed 14-foot-beam boats on to the canal, then the tunnel would have had to accommodate them as well. It is not surprising that he should have found this a daunting prospect. Some commentators have suggested that water supply was his main consideration, but that argument is fallacious. True, a wide boat would use twice as much water to get through a lock as would a narrow-boat – but if you have to divide the cargo between two boats then you will need the same quantity of water, since the narrow lock would have to be used twice. It seems altogether likely that the tunnelling problems were foremost in Brindley's mind. In the event he was justified.

Narrow-boats at Preston Brook, where the Bridgewater and Trent & Mersey Canals meet. The boatwoman at the tiller is wearing a traditional bonnet.

Although in 1767 Brindley cheerfully announced that the work would be completed within five years and offered to accept a £200 wager with anyone who doubted his word; the enterprise was, in fact, to take 11 years to complete, by which time the great engineer himself was dead. Quite simply the technology did not exist to cope with civil engineering on this scale. Everything was kept to the minimum standards. No extra width was allowed for a towpath, so that in the early engine-less days boats were legged through the tunnels, the boatmen 'walking' the boat along by pushing with their feet against the tunnel sides. This was a task made doubly difficult by the fact that only part of the tunnel had a brick lining – elsewhere it was rough, uneven bare rock. The tunnel was also like the Worsley mines, in that side branches led off towards coal faces. Harecastle tunnel took Brindley to the edge of the engineering capabilities of the 1760s – and if that was the best he could do here, then there was no point in going for higher standards anywhere else on the Trent & Mersey. And if that was the standard to be set in one place, then it would only be sensible to carry it through to the connecting links. So a whole system of canals based on boats approximately 70 feet long by 7 feet beam came to dominate canal-carrying in the English Midlands.

A large part of the fascination of canal travel derives from the simple fact that they are wholly artificial constructions. The shapes and levels of the natural landscape may limit the options open to the engineer, but in the end the route taken – the placing of locks, bridges and aqueducts, the design of structures along the way – is the result of decisions taken by individuals, and among those individuals the chief engineer reigned supreme. A Brindley canal is as easily recognizable as his work as a Constable painting is identifiable as that artist's creation. Brindley set his mark on the engineering long before the first spade was pushed into the ground. He favoured the technique known as contour-cutting, where he laid out his canal to follow the natural contours of the land. This had many advantages, one of which being that it removed the need for difficult and expensive earthworks. Faced by a hill across his track, Brindley simply went round it, rather than cutting his way through it. His predilection for avoiding trouble must have made the necessity of digging the long Harecastle tunnel all the more galling. In the valleys, his canals would hug the edges rather than stride boldly across on embankment or aqueduct. He used river valleys to good effect, drawing on the natural streams that fed the river to supply his own canal. He also pointed out that his meandering canals were able to call in at more towns, villages and hamlets than a canal running arrow-straight across the landscape – and one of the features of a Brindley canal was the large number of wharves along the way. Just how much trade these country wharves brought to the canal is not easy to calculate – one suspects it was comparatively small.

The effect of contour-cutting can be seen very dramatically on one of these canals of 1766, the Staffs & Worcester, as it makes its way between Wightwick Bridge at the edge of Wolverhampton and the junction of the canal with the River Severn. As the crow flies, the distance between the two is roughly 12 miles. But Brindley was no crow. His wavering, wandering canal takes 22½ miles to join the two points. The map shows very clearly how closely the turns and wriggles of the canal match the turns and wriggles of the River Stour. A modern Ordnance Survey map also shows the way in which Brindley viewed the land and took his decisions. Kidderminster is a mere two and a half miles from the River Severn and the town of Bewdley, which in the middle of the eighteenth century was a thriving inland port and manufacturing centre. A popular story has it that the river traders of Bewdley scorned Brindley and his 'stinking ditch' and wanted nothing to do with it: but in reality they petitioned for the canal to come to the town. They were well aware that this was bound to be an important junction, where loads were exchanged between the wide vessels of the river and the narrow-boats of the canal, and they wanted their share of the prosperity that the extra trade would bring. But the map shows clusters of tightly packed contour lines between Bewdley and Kidderminster, in the area known as Seven Hills. Brindley looked at this unpromising landscape and turned away, deaf to the pleas from Bewdley. He took his canal on down the river valley to join the Severn at a spot which was then marked by little more than a country tavern.

The effects of this decision were exactly what the citizens of Bewdley had feared. Their trade began to slump as more and more vessels ended their journeys at what was rapidly to develop into the new town of Stourport. Here one can still see, as clearly as anywhere on the whole system, both the importance of the canal and the effect of Brindley's decision to opt for the narrow lock. There are two ways of reaching the basins beside the river – through narrow locks or via the 15-foot-wide broad locks. Here cargoes could be exchanged directly between river and canal boats, or unloaded into warehouses to await transhipment. At once this created a demand for warehousemen and wharf managers, clerks and dockers. The new port became a place where men could trade and deal, and so around the dock complex a whole new town was developed. One of the features that becomes particularly apparent at Stourport is that there is no real distinction being made stylistically between commercial and domestic architecture: both adopt the mannerisms and rules that governed Georgian building and which give it its unique air of cool elegance. You could take a warehouse window out and pop it into a terrace house and not be aware of any incongruity. The Tontine Hotel is, if anything, rather less attractive than the warehouse with its delightful clock tower. The appeal of Stourport is immediate and obvious, and is enhanced by the nature of the waterway system itself: reflections add a shimmering image of the buildings; the hard angularity of black-and-white balance beams and iron paddle gear contrasts with the well-worn richness of old brick walls.

The second of the two pioneer trunk routes, the Trent & Mersey, resulted in a similar new town development, but in rather different circumstances. Again the narrow canal had to meet the wide waters of a navigable river. Here, however, there was no problem about the selection of a junction, for canal and river kept company for many miles until the limit of navigation was reached at a point on the Trent upstream from Nottingham. Again a Georgian new town, Shardlow, grew up at the junction. Here, however, instead of being grouped around basins, there was a linear development along the length of the canal. Consequently one does not have quite the same sense of looking at a closed, self-contained group of buildings. The buildings are taken sequentially, so that the eye is drawn to details rather than the panoramic scene. One notices the simple use of cast iron for bridge numbers and mile-posts – the mile-posts are, incidentally, of a pattern also used on the area's new turnpike roads. Warehouses, too, have partly borrowed their language from road-side buildings. Next to the two pubs, where generations of boatmen have slaked their not inconsiderable thirst, is a red-brick warehouse which at its lower level has a rounded corner, but above which it is corbelled out so that the walls meet at the conventional right-angle. This is visually appealing, but also wholly practical. The warehouse provided an interchange between land and river transport, and waggons that bump a rounded corner will slide round it, whereas they would damage a straight-edged one. Time and again one finds in the old buildings this happy correlation between the functional and the visually appealing. The best known of all the Shardlow buildings is a case in point. The former Trent Mill is notable for the graceful arch on which it stands. This was built to allow boats to be steered in underneath the main building, so that they could be loaded and unloaded in the dry. Similar covered loading bays can be seen in warehouses throughout the system. And there are variations on the theme. On another early canal, the Chesterfield, the old Pickfords depot at Worksop was built not over a short arm as at Trent Mill, but over an arch that crosses the main canal. This wholly practical building is perfectly designed for efficient use. Boats pull in under one half of the buildings; carts are loaded and unloaded through waggon doors that face out on to the wharf.

This is a crucial element in the understanding and appreciation of canals. Built for use, they still manage to be systems which, generations after they were completed, seem wholly satisfactory as elements in the landscape. Indeed, if this were not so, there would never have been a movement to preserve and keep them. Why should this eighteenth-century transport system have such a hold on the imagination of those who travel that same system for pleasure in the twentieth century? To understand that, one has to look at the circumstances in which they were built.

In the late eighteenth century, Britain was still a small country, in the sense that most of its inhabitants seldom strayed more than a few miles from the place in which they were born. But it was not just people that never travelled. The canals came into being because road transport was inefficient, expensive and unreliable. So, as far as possible, people drew on what was available close at hand. In the heavy clays of the Midlands, they burned these clays to make bricks; in the rough, stony country of the Pennines, they used the stone to build their houses. The canal system was destined to help the process of change, where slate from North Wales would provide a durable alternative to thatch from local reed beds; where mass-produced bricks, standard in colour as well as size, would introduce a new uniformity to streets in towns and cities. But during the construction of the canals themselves the old ways still prevailed. The canal-builders chose the materials that were locally

available, so that what we see today are structures that have an intimate relationship with the ground on which they stand. An excellent example can be found on the southern section of the Oxford Canal, begun under Brindley's direction in 1769.

The southern Oxford shows a rich variety of elements in its construction: elements that reflect both the conditions imposed by the landscape through which it had to pass and the changing fortunes of the canal company itself. Starting with the natural landscape, what one finds is a canal that epitomizes the Brindley philosophy. It begins by following river valleys as closely as possible. At first this means the flat meadows that stretch out from the River Thames as it makes its somewhat devious way round the city of Oxford. It has hardly got under way, however, before the Thames is abandoned for its tributary, the Cherwell. It is easier to see the relationship between river and canal on a map than it is when travelling the waterway. Each twist of the natural river is followed by a turn of the canal, and as the river is shadowed upstream, so too the canal changes level by a steady progression of narrow locks. But Brindley's enthusiasm for river valleys was not only roused by the comparative ease with which the canal could be fitted to the landscape; he also saw the river as a useful water supply. Between Enslow and Shipton-on-Cherwell, the canal boats leave their calm waters for a brisk excursion on the Cherwell itself. The difference in level between river and canal is quite small, with a drop at Shipton of only 2 feet 5 inches. On a canal this would not matter, as the excess water would simply pass over a by-weir into the next pound. Here, however, water not going through the lock would continue on its way down the Cherwell. So a conventional lock would have resulted in a rapid drying up of the lower sections, and as there was nothing to be done about depth, the lock was built in a curious diamond shape, so that the increased width would ensure a good volume of water passing through.

North of Banbury, however, the river reduces to the proportions of a stream, and river and canal are forced to diverge. From the top of Claydon locks to Napton is a landscape that might delight the eye of an artist, with its lovely green irregularity of hills and hummocks, but which promises only headaches and problems to the canal engineer. It was decided to keep to just one level, while at the same time restricting engineering works to a minimum. At Fenny Compton, the ridge had to be pierced by what was originally a shallow tunnel. This was an unnecessary inconvenience and in 1838 the land above the tunnel was bought from Christ Church, Oxford, and the top of the tunnel was removed, creating a deep cutting. Beyond that, however, the canal proceeds through a bewildering series of twists and turns. Travelling it by boat one sometimes sees another canal alongside – except that it is not another canal at all, simply the Oxford going through one of its many about-turns. The most dramatic example can be seen at Wormleighton, where the canal

Kidderminster, Staffs & Worcester.

initially encircles the hill. You pass the front of New House Farm and then almost half an hour later you chug past the back door. All this convoluted travel comes to an end when the watershed has been safely crossed and the canal, thankfully it seems, dives down through the Napton locks, with one last squirm around Napton hill before continuing on its way north to its eventual junction with the Coventry.

Jumping ahead slightly in time, it is interesting to take a look at the northern section of the canal. This was shortened in the 1820s as an answer to the threat of competition posed by a proposed broad new canal, to be called the London & Birmingham Junction. The Oxford proprietors called in Marc Isambard Brunel, famous father of an even more famous son, to survey a new route. By cutting straight lines across the curves, he was able to make a saving of over 13 miles. Nothing gives a clearer idea of just how meandering the old canal was than a comparison with the new. If you leave the new canal at Newbold wharf by the tunnel and walk down to the church, you will discover a second tunnel – narrow, low and completely dry – set virtually at right-angles to the present canal. If you then follow the dip in the land that marks the former canal bed, you come across little crumbling red-brick bridges set incongruously in the middle of fields and even the remains of long-abandoned, overgrown wharves. Slowly your trail will take you round in a great semi-circle until you are back once again at the new canal. It gives a very clear idea of how canal technology and, just as important, canal thinking had changed in just half a century.

The original Brindley line of the Oxford Canal at Newbold on Avon, with the old tunnel. The boat has no back cabin.

The first half of the canal to be built had brought it as far south as Banbury, but by the time the first boats reached Banbury in 1778, James Brindley was dead and the work had been taken over by his assistant, Samuel Simcock. More importantly for its future development, the company had run out of money. A new Act of Parliament was needed and all work ground to a halt until 1787. Oxford was finally reached on 1 January 1790 amid great celebrations as a fleet of boats arrived loaded with over 200 tons, mostly comprising coal and corn. Although the canal's construction covered such a long period of time, there is a remarkable homogeneity about the structures that are met along the way. In general, brick rules for the familiar hump-backed bridges, but what a beautiful brick it is, rich in tone and texture, reflecting the richness of the countryside through which the canal passes. It does, however, also pass through a band of oolite, north of Kidlington – the importance of the local stone quarries is reflected in the name of the canalside pub, the Rock of Gibraltar. With such good building material readily to hand, brick gives way – if only temporarily – to this attractive pale limestone. Availability ruled decision-making, but as a result the canal structures not only seem a part of the natural landscape, but are directly related to its underlying foundations. This applies to more substantial buildings (lock cottages, warehouses, wharves) as well as to the ubiquitous bridges.

But the financial constraints of later years brought a new element into the landscape. The traditional bridge rises steeply over the canal, but even so still requires a good deal of material to lift it high enough to allow boats to pass underneath. It is far more economical of materials to bring road or path up to the water and carry it across on the level. This can be done, provided that the bridge can be moved to allow boats through. The result can be seen in the many simple lift bridges. They are absolutely basic, with the weight of the bridge platform balanced by massive beams rising up at an angle – rather like a lock gate set on its side. This also represents a financial balancing act, for while the moveable bridges are unquestionably cheaper to make, they are more expensive to maintain. For a cash-hungry canal company, anxious to complete its canal to get the revenue flowing, the short-term gain far outweighed the long-term costs.

The visual effect of a canal such as the Oxford, or more particularly the southern Oxford, derives even today from decisions taken when work began more than two centuries ago. Yet there have inevitably been great changes. As the carrying days grew to a close, so wharves became abandoned, warehouses redundant. Some found new uses, such as the conversion of part of the warehouse complex in the heart of Banbury – and, seeing how successful that is, one can only mourn the destruction of so many of the other canalside buildings in the heart of the town. Brindley would have been amazed by, but would undoubtedly have approved, the latest transport system to push its way up the Cherwell valley. Cars thunder over the formerly peaceful canal on the new M40, destroying the calm for ever. But Brindley built for trade, not for aesthetics. When Josiah Wedgwood, principal promoter of the Trent & Mersey, planned his own home, Etruria Hall, beside the canal, he begged Brindley to give him a graceful curve to set off his lawns. But as Wedgwood complained in a letter to a friend, the 'inflexible vandal' would not allow an inch of deviation from the line he had devised, however hard Wedgwood might plead for his 'line of grace'.

The delightful meanderings, the mellow brick, the simple vernacular style of buildings, which so many of us admire today, have no basis in any aesthetic theory, but are practical responses to practical problems. If Brindley had had access to concrete and steel he would have used them with pleasure. These early canals were, indeed, carriers of change, but they did not themselves partake of it. Cottages were not so much conscious designs as simple continuations of local traditions, and generally the work of a local builder. The changes that the canal brought can be seen in the growth of industry and settlements along the banks, while the waterway itself remains a strip taken out of time. As such, it is a world within a world, which carries one back to a time when local styles were not provided at the whim of an architect or builder, but were forced by circumstances.

The appeal of the canals of the Brindley age lies very much in their 'natural' line, their simple unaffected building styles and the happy accumulation of small details. Yet one must never lose sight of the fact that this visual appeal was not the main concern of the canal-builders. To them, canals were about efficient transport and fat profits. It was the growth of industries along the banks that were the measure of their success. Wedgwood might have complained about the view from his house, but his real concern was the mighty pot-works that he built at Etruria to be served by the canal. There was an intimate relationship between the pottery and the waterway devised to serve it. The potteries areas had developed around two basic commodities: good clay, suitable for earthenware, and coal for the kilns in which the pots were fired. But taste was changing in eighteenth-century England. The robust, heavy designs of one age were considered too crude for a new age that had discovered the delicacies of oriental porcelain. The Stoke potters could not use the processes of China directly, but they could produce a good approximation by importing materials such as china clay and flints, which could be mixed with local clays to make a lighter and more delicate ware. But to make the new processes pay, transport costs for these bulky materials had to be reduced – hence the enthusiasm for canals. There was an added bonus in that the delicate finished pots could be crated and sent by water, instead of being sent over bumpy roads by waggon or pack-horse.

Caen hill locks, Kennet & Avon.

The effect of the arrival of the canals can be seen throughout the potteries. New works scrabbled to get a waterside site; distinctive, shapely bottle kilns rose up above the placid waters. Here, at least, the connection between canal and industry is plain for all to see. In other areas the link is a good deal less obvious. Burton-on-Trent also stands close to the canal, a town with its own particular industrial associations. Brewing began in the area in medieval times, but the coming of the canal helped a purely local concern to become national and, ultimately, international. William Bass was, in the middle of the eighteenth century, a man who combined a little brewing with work as a carrier. He found the beer trade more to his liking than carrying and came across another local who was anxious to expand in the transport business. Mr Bass sold his carrying interest to Mr Pickford, and the combination worked so well that in time Bass were able to maintain their very own warehouse at Paddington basin, stocked by regular deliveries, courtesy of Pickfords' boats. Bass was always on the look out for new ideas, and when the Bar Brewery in London developed a light, sparkling beer with a distinctive bitter tang, Bass started producing the tasty brew and found a market for it among the thirsty servants of the Empire. It became known as East India Pale Ale, later simply IPA, and it began its long journey to the Orient along the Trent & Mersey Canal.

Other canals can seem quite divorced from industry. Today the Ashby offers a wonderfully relaxing, lock-free meander through peaceful countryside. Those who come this way by boat get only the most obtuse hints as to why it was ever built. But the canal now is shorter than it was when first built. Continue the journey, not by boat but on foot along the towpath, and you arrive at the town of Measham, once famous for its pottery. The local ware was known variously as Meashamware, Rockingham and also, because of its popularity with boating families, as bargeware. Further along the canal is Moira, where an immense blast-furnace was built into the canal bank, with the loading bridge actually spanning the waterway. The pretty, rural canal turns out not to be so rural after all.

Nowhere is the intimate relationship between burgeoning industry and the spread of canals more dramatically illustrated than in Birmingham. In the eighteenth century, waterways were compared to the veins of the natural body, feeding and stimulating growth. This was certainly true of Birmingham. Industrialists rushed to acquire waterside sites, where coal, the fuel of the Industrial Revolution, would be readily and cheaply available. Great factories such as that of Boulton and Watt, the pioneers of steam-engine production, were wholly dependent on the boats that served them. And around the great factories, a myriad of small workshops clustered. Birmingham was canal city – and could scarcely have developed without them. Yet the canals that initially brought prosperity were those of the first generation – notably

An island toll-house at Smethwick on the Birmingham Canal. Boats stopped in the 'narrows' to be gauged, so that tolls could be calculated.

the typically meandering Birmingham Canal itself which, like the northern Oxford, was to be supplemented by a new, direct main line, leaving the old curves as loops and byways. It takes an act of historical imagination to understand the significance of Birmingham's canals. The Birmingham Canal is today overshadowed by railway and motorway, yet this is the improved canal of the nineteenth century. The old, looping canal remains as an oily, stagnant ditch, home to dead cats and supermarket trolleys. This backwater, though, is the one that helped create the canal mania. It was shares in this modest canal that shot up in value; it was the trade on this now all-too-modest waterway that laid the foundation of Birmingham's prosperity. Even before the mania years, a spidery web was beginning to form around the original Birmingham Canal – with the Birmingham & Fazeley, the Dudley and the Stourbridge Canals.

Take it out of its urban context and you can still see the Birmingham Canal as one of the family of Brindley canals that snaked their way across the landscape. Brindley may have dominated the early years, but his was not the only voice to be heard. Other engineers went to work producing schemes of an equal grandeur – and, in some cases, outstripped Brindley in the breadth of their ideas. By far the most ambitious was the Leeds & Liverpool, a coast-to-coast canal that had to find a way through, round or over the Pennine hills. It was not merely ambitious, but over-ambitious. John Longbotham went to view the Bridgewater Canal and, on the strength of that visit, blithely asserted that he saw no reason why a broad canal, 127 miles long, should not be built on the same principles. In terms of practical engineering he was proved right – but in terms of finance his estimates were to prove sadly awry. The first Act was passed in 1770, but by 1777 the money was spent. Other Acts, empowering the company to raise more funds, were passed in 1783, 1790 and 1794, with a grand opening of the completed work in 1816 – scarcely a spectacular rate of progress. It all looks, from the bare facts, rather pathetic, but in that first hectic burst of activity works were completed that quite overshadowed anything being done on the Brindley network of the Midlands.

To get some idea of what was achieved one could look at the solutions provided to a similar problem: how to overcome a sudden steep change of level. Brindley met that challenge at The Bratch on the Staffs & Worcester. His solution was bizarre. Looking at the locks, they present something of a mystery: three locks set so close together that there is less than a boat's length between them. This does not present a problem as far as using the locks is concerned, even if it is a trifle unusual, but it does represent a problem in water supply. A boat coming down would, it seems, push out a lockful of water that would more than fill the next pound, and in a conventional system the overflow would simply disappear over a weir, so that there would not be enough left to fill the next lock. Close inspection reveals the answer – side ponds lurking among the undergrowth to act as mini-reservoirs. The pounds may be short, but they are given this extra width. The Bratch has become something of a canal showpiece. A neat little bridge hops over the middle of the flight, with a minimum of fuss and with great economy of material: indeed, it has been taken down so close to the water that a recess had to be built in the abutments to allow for the swing of the balance beam. At the opposite end of the bridge stands a handsome octagonal toll-house, with round-headed windows providing a perfect view of all passing traffic. But why did Brindley adopt this very cumbersome method of working the locks? The obvious alternative can be seen in an even more spectacular site on the Leeds & Liverpool.

Work on the canal began simultaneously at both the Leeds and Liverpool ends. At the Leeds end, the obvious course was to follow the line of the Aire valley. This has few of the extravagant curves that Brindley so faithfully followed as his canals duplicated the line of the Cherwell or Stour, but it does have a far steeper gradient, necessitating a large number of locks. The canal leaves the river at Leeds itself and eventually arrives high in the hills, where it enjoys the luxury of a 17-mile section quite free of locks. But to get to that happy position it has been raised by 240 feet.

On a Brindley canal one would have expected to find a scatter of locks all the way down the canal, but not here. They come in groups, separated by long pounds, and not just grouped together in flights but run together in staircases. In this system, the lock gates open directly into the next lock with no intervening pound. For a boat travelling downhill, this presents no problems: the top lock fills the next one down, and when the levels are equal the gates are opened, the boat moves to the next lock, and so on for as many steps as there are in the staircase. Going uphill is more complex, as the lock above has always to be full, so that there is a constant feeding of water down through the locks. The instructions make alarming reading, but in practice common sense makes the system clear. Which is just as well, for there are eight staircases in the first 17 miles, reaching a grand climax at the end, first in two locks together, then three, then the last giant leap of 60 feet at the famous Bingley Five rise. And these are not the narrow locks of the Midland system, but broad locks to take vessels 14 feet 3 inches wide by 60 feet long. Bingley is a majestic sight with its array of black-and-white balance beams and its stern, stone cottage marking the arrival at the top. This is civil engineering on a truly grand scale and time and again the Leeds & Liverpool shows a grandeur which matches that of the countryside through which it passes.

This is not the only spectacular feature that this section of the canal has to offer. The area through which it journeys was already well established as a centre of the woollen industry, but that industry was still based on the age-old system of spinning in the cottage and hand-loom weaving. Only the fulling, or finishing, process took place in water-powered mills. Change, however, was very much in the air. First spinning, then weaving, moved from home to factory, and over the years power changed from water-wheel to steam engine. The canal was to play its part in this great movement as the principal means of transport for raw materials and finished products – indeed, without water transport, the whole progress of the Industrial Revolution would have been a good deal slower. This is obvious today in the mills that line the route – some, happily, still at work; others abandoned or converted to new uses. Nowhere, however, is there a clearer indication of the importance of water transport to the industrialist than at Saltaire.

Sir Titus Salt was a manufacturer of mohair, and from 1848–9 he was Mayor of Bradford. He looked with horror on the conditions in which the woollen workers lived and was determined to provide a shining example to all his fellow-manufacturers. A splendid new Italianate mill was built, straddling the canal. Just as important was the town he built to go with it. Houses followed the same Italianate theme – neat villas sat beside broad streets. Almost every facility was provided for the well-being of the families – library, hospital, church, almshouses, a park. But there was one notable omission: the teetotal Sir Titus made sure that no public houses invaded his Utopian settlement. It was a forerunner of the better-known industrial villages, such as Bournville and Port Sunlight, and is still quietly attractive, though the machinery of the mills is now silent. Sir Titus Salt's town by the Aire is a mill town first, but it is also very much a canal town as well.

Throughout its length, the Leeds & Liverpool Canal shows a dual personality: at some points passing through magnificent Pennine scenery; at others running through a stone canyon of mills that rise sheer from the water. Although construction covered a very long period, there is no obvious difference between the early sections and those completed at a much later date. The Pennine section has all the qualities that one associates with an early canal. The hills force it into extravagant convolutions and there is the same reliance on local materials, which in this region means stone. Look, for example, at the section surrounding the locks at Greenberfield. There is a wonderfully satisfying sense of unity about the place. The canal rounds the flank of a high, smooth grassy hill that overlooks a landscape squared off by dry-stone walls. The lock cottage, built of the same dark stone, is very much in the Pennine vernacular: smooth slabs of lintels contrast with the rough-hewn blocks of the walls, and the stone mullioned windows add to the air of solidity. A sundial on the wall provides the only decoration, and none other is needed, for the rich texture of stone, the subtle gradations of colour, light and shade provide all the visual interest that the façade requires. The same satisfactory simplicity extends to everything on and around the canal. It is bounded by the same stone walls that divide up the fields; the stone bridges are gently unobtrusive and the wooden footbridges are perhaps the neatest, most economical features of all. A single baulk of timber provides adequate footing, and hand-rails splay out to allow for the fact that all of us (some to a lesser, some alas to a much greater, extent) spread outwards somewhere around our middles.

If I were to take a stranger to see the Leeds & Liverpool at its best, in a rural Pennine setting, then Greenberfield would be my choice; if I wanted to show the same canal in its working role, among the textile mills, then Burnley would be the first stop. The same stone reappears, but now on a monumental scale. Yet the architectural language is also largely unchanged. The regular façade of windows, with their slab stone lintels, represents little more than the cottage or house writ large. Some details are particular to the new environment, like the canopied loading bays, but in the early years of the Industrial Revolution the change from cottage to factory appears as a change in scale, but not a change in style. The change in scale appears as the canal moves away from the close, confined area of the mills to cross the Calder valley on a high embankment. Here one has reached the later years of

Warehouses and cotton mills line the Leeds & Liverpool Canal on the approach to Burnley.

the canal-building programme, and one is faced by a section of canal which speaks loudly and clearly of the second phase of development.

If the Burnley embankment proclaims itself as unmistakably belonging to the later years of canal construction, this does not mean that everything that happened in the early years was somewhat timid and small-scale. In Scotland, the canal age got under way in great style. The Forth & Clyde Canal was conceived, from the first, on a grand scale, with locks 74 feet long by 20 feet wide and, just as important, able to take craft of 7 feet draught – later increased to 10 feet. The idea of a waterway route across Scotland was first mooted in 1768, but nothing of any consequence occurred until that eminent, pioneering engineer John Smeaton travelled north to cast his expert eye over the project. He was only one of a number of engineers who were approached to give their views, and among the others was that doyen of the early canal age, James Brindley. Brindley was a man very much accustomed to getting his own way, but in John Smeaton he met his match. Brindley was by now hopping around the country, busying himself on a variety of schemes, and he raised objections to the project. Smeaton declared there was only one true objection, and that he proceeded to demolish with wholehearted ruthlessness. Brindley mentioned that he was worried on 'many accounts', to which Smeaton replied:

I wish you had been a little more explicit on the many accounts: I think you only mention one, and that is to give more time to examine the two ends: but pray, Mr Brindley, if you were in a hurry, and the weather happened to be bad, so that you could not satisfy yourself concerning them, are the works to be immediately stopped when you blow the whistle, till you can come again, and make a more mature examination?

At this point Brindley retired hurt, and the work went ahead to Smeaton's plans. At first, everything went smoothly, but having fallen out with Brindley, Smeaton now fell out with the company and, without his controlling hand, work came to a halt. Funds ran out and there was to be a nine-year gap before work was resumed under Robert Whitworth and the canal was opened in 1790. This was canal-building on a truly grand scale and, with an array of moveable bridges along its length, the Forth & Clyde could be used by vessels of considerable size. The facilities were on an equally grand scale, notably Port Dundas at the Glasgow end, with its imposingly classical headquarters. Structures were not so much grand as monumental, including the noble four-arched Kelvin aqueduct. It is ironical that this magnificent canal, which was to prove itself a huge success – not only in returns in investment, but in the industries that were attracted to its banks – should have faltered for lack of funding. It might never have been completed at all, had the government not stepped in to help with the finance. Sadly, it was a lack of government imagination that was to bring about its closure. Again, irony is in evidence. The modern age of canal traffic is dominated by pleasure-boating, and the Forth & Clyde was among the pioneers of passenger trade. As early as 1783 it was running boats carrying goods and people – boats celebrated in verse by James Maxwell, a poet who, not surprisingly, has slid into obscurity:

> For here a cabin in each end is found,
> That doth with all convenience abound,
> One in the head, for ladies nine or ten,
> Another in the stern, for gentlemen,
> With fires and tables, seats to sit at ease;
> They may regale themselves with what they please.
> For all utensils here are at command,
> To eat and drink whate'er they have at hand.

Not content with inaugurating a regular passenger trade, the Forth & Clyde company was among the first to experiment with steam on the canals. The steam-tug *Charlotte Dundas* began a series of trials in 1801. The stern-wheeler, designed by the American engineer Robert Symington, was highly successful in its task of hauling barges, but because it was felt that the wash would damage the banks, the project was

Bradford-on-Avon, Kennet & Avon.

dropped and the paddle-tug left to rot and decay. Steamers were to use the canal in later years: vessels as varied as the little all-purpose coasters, the Clyde Puffers and imposing pleasure-steamers. With such a rich and innovative past behind it, the closure of the Forth & Clyde seems especially sad. It could be said to have brought a vision of the future to the world of canals – a vision of powered boats and pleasure-cruising, of wide commercial waterways that would link great cities across the country. It stands as an achievement that ranks as high as anything of the first canal age south of the border in England: though works were in progress there that were far from negligible.

One of the most ambitious canal schemes was that for uniting the Thames and the Severn rivers. A start was made in 1779 when work began on the Stroudwater Canal, which was to link the Severn at Framilode to Stroud, then the centre of a thriving woollen industry. It is difficult today to think of this region at the edge of the Cotswolds as an industrial area at all, but the local rivers supported literally hundreds of woollen mills, large and small. As on the Leeds & Liverpool, the arrival of the canal encouraged the building of new and ever-grander mills. Stanley Mill at King's Stanley is a wonderful example of Georgian mill architecture, the outside enlivened by such attractive features as giant Venetian windows. But the finest effects are reserved for the interior, where the iron columns that support the upper floors are arranged in surprisingly delicate and ornate colonnades. The ironwork was cast in Dudley and brought down by river and along the new canal. Nearer to Stroud centre is the even grander Ebley Mill, built somewhat in the style of a French château, which has recently undergone a somewhat surprising transformation into council offices. So the canal was seen from the first as an important boost to local trade, and soon more ambitious plans were being considered, for an extension to the Thames at Lechlade.

Here a major obstacle presented itself. The Stroudwater had climbed steadily up the valley of the Frome, but the valley turned north towards the river's source high in the Cotswolds and the canal had to carry on towards the east. There was no way over the top, so it had to be pushed through. In 1766, Brindley's concern over building the Harecastle tunnel had been an important factor in his decision to opt for a narrow canal. The Stroudwater Canal had been built to take river vessels up to 70 feet long and 15 feet 6 inches beam: this proved a little too daunting, but the decision was taken to build a barge canal to take vessels up to 11 feet beam. The small-bore Harecastle tunnel had taken 11 years to complete; Sapperton tunnel, broader and 3,817 yards long, was to be completed in just five years. The going was far from easy: sometimes the builders met solid rock that had to be blasted away; sometimes difficult-to-work fuller's earth was encountered; and springs caused perpetual drainage problems. A series of shafts, mostly over 200 feet deep, were sunk down to the level of the tunnel at regular intervals, and from the bottom of these the gangs worked outwards. The work went on day and night, and the labouring gangs were paid at the rate of £5 a yard, out of which they had to find their own materials. It remains a mightily impressive work.

At the Coates end, the tunnel portal was given the full classical treatment, with prominent voussoirs, attached columns and niches. The other end was more modest, with simple castellations. Yet the true measure of the achievement becomes apparent only when you go inside. At first, it is no more than a brick-lined tube, but soon the lining stops and the bare rock appears. Overhead are great flat slabs, at either side broken and shattered by blasting. Semi-circular grooves show where the men drilled holes that were packed with black powder. It is not hard to imagine the scene, lit only by candles that scarcely penetrated the clouds of dust that hovered after the repeated explosions; nor is it difficult to imagine the hard labour in the dripping dark, as the navvies cleared away the loosened rock with pickaxe and shovel. One thing that is difficult to imagine, however, is how the boats were brought through. There is no towpath, and the rough, uneven walls would have made legging very difficult, if not impossible, as the width is far from even. Whatever the answer to that question, one thing is clear – the creation of Sapperton tunnel was an achievement of epic proportions.

The whole canal is full of interest and has a number of quirky features, including round lock cottages. There seems to be no documentation to explain why the design

The Thames & Severn has unusual, round lock cottages: this one is at Lechlade.

was adopted, but it is perhaps relevant to note that they would have looked less outlandish when the canal was new. Then similar circular towers were a common feature in the landscape. They were wool-drying stores, but one at least has been converted into a house and the resemblance to the lock cottages is certainly striking. Other features that were once equally interesting have been lost since the canal was closed. An inland port developed at Brimscombe near the junction of the Stroudwater and the Thames & Severn, with a substantial basin and a range of stone-built warehouses. Once the canal was open, carriers such as Edmund Wells of Lechlade were able to offer a service between Bristol and London, using Severn boats to the west of Brimscombe and Thames boats for the rest of the journey. Now the basin has been filled in and the buildings demolished. What has not changed is the superb scenery through which the canal passes, so that its eventual restoration is something to which all enthusiasts can look forward.

The first canal age had seen nearly all the techniques developed that were to carry canal-building through to the end of the century. River valleys had been crossed by aqueducts and embankments, hills pierced by tunnels and cuttings. Works such as Sapperton tunnel had shown that the timidity of the early years had been left behind. Locks could be narrow or wide; they could appear singly, spread out evenly over the whole length of a canal, or clustered together in long flights and even joined to create staircases; and if none of these alternatives appealed, the canal could always wander off to find an easier, level route around the obstacles. Inevitably, however, there came a time when engineers wanted to build canals in a landscape where the hills were simply too steep to be conquered by locks, too broad to be pierced by tunnel and too numerous to be circumvented. In such circumstances either canal-building had to be abandoned or new solutions to the problem had to be found. In the end, ingenious men found an ingenious answer. In 1788 the Shropshire iron master, William Reynolds, constructed England's first inclined plane on the Ketley Canal.

At the top of the plane was a lock. A boat was floated in and, as the water was let out, so the boat sank down until it settled on a wheeled wooden platform. This stood above a railed track, enabling platform and boat to be lowered under their own weight to the bottom of the plane and a second section of canal. This was coupled to a second track, so that the weight of the loaded boat going down the plane could be used to raise a second boat up it. This system was to find widespread application, notably on the Bude Canal, where the great Hobbacott Down plane had a vertical lift of 225 feet. Although the planes on the Bude Canal can still be traced, no mechanism survives. Another, much later and equally impressive plane, can be seen at Foxton on the Leicester arm of the Grand Union. But for a clear idea of how the system worked, the best place to go is Coalport, where the plane that links the upper level of the canal on the Blists Hill museum site to the lower level near the Coalport Pottery on the banks of the Severn has been restored.

By the time canal mania reached its peak, several things were clear. The basic problems of canal-building had been solved. Those who had scoffed at Brindley and the Duke of Bridgewater had long since been silenced. The canals were a success. They were a success in meeting the needs of industrialists, who were demanding regular supplies of raw materials at reasonable prices. They were a success in promoting development along their banks. There were canal corridors, much as today we have motorway corridors. Above all, they were a success in that the best of them gave investors a more than handsome return on their capital. What was not so immediately obvious at the beginning of the 1790s was that the canal world was about to enter a new phase. Engineering would become more sophisticated; there would be a new bravura to characterize the later days of canal development. What was equally unforeseen, if not necessarily unforeseeable, was that the end of the canal age was little more than a quarter of a century away. Yet all the developments that were to culminate in the move from waterways to railways were already present, in embryo at least, in the canals of 1793.

Few early canal tunnels had towpaths, so boats were legged through. These men are coming through Butterley tunnel on the Cromford Canal.

NARROW-BOAT COUNTRY

If any canal could be said to typify the new age of canal construction in the mania years, then that canal is the Grand Junction. By the 1790s, the shortcomings of the first generation of canals were already becoming clear. The Industrial Revolution was gathering strength with every passing year, and as it did so trade inevitably increased. Boats travelling between Birmingham and London began their journey somewhat unpromisingly by heading north-east on the Birmingham & Fazeley to the Coventry Canal, and then on to the Oxford. For a time, a farcical situation occurred whereby the Oxford and Coventry Canal companies could not reach an agreement on where they should meet, so that the two waterways ran side by side for many miles. Eventually, this was resolved and a new junction created at Hawkesbury – known to generations of boat people as Sutton Stop. Having joined the Oxford Canal, the boats then proceeded to wiggle and waggle through Brindley's convolutions until the broad valley of the Thames was reached. Even then their troubles were far from over, as the river takes an even less direct course than anything Mr Brindley would ever have countenanced. The case for improvement was clear.

The Oxford Canal Company was well aware that something had to be done – and was understandably nervous that whatever was done would strip it of its trade. It responded by promoting a new canal, the London & Western or the Hampton Gay Canal. Today Hampton Gay consists of no more than a ruined manor house, and a few bumps and dents in the ground where a medieval village once stood. But nearby is the little canal settlement of Thrupp, one of the most delightful spots on the canal, with its waterside pub, and its little terrace of stone cottages – where even the canal-maintenance yard buildings come complete with rural thatch. It was from here that the new canal was to set off across country for London. But this was not what the new age wanted: a narrow canal that still relied on the wavering southern Oxford was not an attractive proposition, and doubts were raised about the

New lock gates being installed on the Grand Union. The long-handled scoops are shown in an illustration of canal workers' tools in the eighteenth century.

practicality of the scheme, particularly with regard to water supply. The project even found itself lampooned in a broadsheet, billed as 'The Canal With No Water'.

The alternative was the Grand Junction, designed from the first as a broad canal, able to take Thames barges and running from the Thames at Brentford to the Oxford Canal at Braunston. The idea was then to widen the remaining section of the Oxford and persuade the other canal companies to do the same on their waterways. That they failed to do so was one of the great mistakes of the canal age – for the old rule still applied. One narrow section in the entire system was all that was needed to ensure that the narrow-boat ruled throughout. There was no real contest between the two proposals, and once the Oxford had been bought off with compensation for loss of trade, the way was open for the Grand Junction Canal.

The first survey was carried out by James Barnes, who had himself worked on the Oxford, and his suggested route was then sent to William Jessop, who approved it, with a few modifications. When the canal was authorized, Barnes was to become resident engineer, working full-time on the Grand Junction, while the job of chief engineer went to Jessop. It is only in recent years that Jessop's status as a canal engineer has received the recognition it deserves. Yet throughout the mania years, his was the dominant influence. Many commentators have leaped from the age of Brindley to what they call the Telford age – yet when Jessop was put in charge of the Grand Junction he was already acknowledged as one of the foremost engineers of the day, with an impressive record of successful waterways work behind him; Telford, on the other hand, had scarcely started his career, had yet to work on any waterways and was, indeed, to get his first experience of canal engineering as Jessop's assistant. Who was the man who was to carry so much of the burden of overseeing the mania years, and why is his name not better known?

William Jessop's father was a foreman shipwright, who became involved in civil engineering when he was given responsibility for the maintenance of the 1715 Eddystone lighthouse. This caught fire and was destroyed in a great gale in 1755, and the job of building a replacement went to one of the greatest engineers of the day, John Smeaton. It was natural that the elder Jessop should be involved in the new work; natural, too, that young William should find himself apprenticed to Smeaton. It could be said that the storm of 1755 was one of the best things that could have happened as far as the boy was concerned – for it gave his practical education over to the hands of a master. From the day when as a 14-year-old boy he took up his apprenticeship until 1772, when he first set out on his own, he worked for Smeaton first as apprentice and then as assistant. He worked on major river navigations such as the Calder & Hebble, and on major canals, including the Forth & Clyde, as well as having the chance to engage in other branches of his trade, including harbour construction

Fishing boat, Caledonian Canal.

and land drainage. Over the next few years Jessop worked his way steadily up the professional ladder, gaining more and more experience as he went. By 1793 he was mature and ready to grasp the opportunities on offer. For the next few years there was scarcely a major canal scheme in Britain for which he was not either personally responsible or on which he was not called in for an opinion. Which brings us back to the other question. Why is this important figure not better known?

Partly this was a matter of personality. Jessop was, it seems, an unassuming, practical man who went about his work with professional competence. He did not have the curiosity value of having come from the most modest of beginnings, as did the semi-literate Brindley or the poor shepherd's son, Telford. Nor did he have the extravagant ego of the heroes of the railway age, such as Brunel. He was modest in his life and, just as important, equally modest in his work. He rarely went in for lavish structures,

Narrow-boats and lighters cluster round a freighter at Limehouse dock.

*The original Grand Junction
Canal office at Paddington,
from a painting of 1820.*

and showed little enthusiasm for architectural embellishment. It is not easy to define a 'Jessop style'. He was a man willing to make use of whatever technique seemed appropriate to the circumstances. His strong point, as Charles Hadfield and A.W. Skempton made clear in their biography, was the basic one of managing the elements of earth and water. He was a man willing, sometimes it seems almost eager, to own up to mistakes. There was never any attempt to shift the blame on to others, even though in the hectic years of the 1790s he could not possibly give all his time to the different projects in which he was involved. He had to rely on assistants who were not, alas, always reliable. The Grand Junction shows very clearly both Jessop's strengths and the problems that he faced – and which he was not always able to overcome.

The line of the canal was substantially that laid out by Barnes, though at the London end the junction was moved to the outfall of the River Brent, and Jessop suggested a straightening near Leighton Buzzard. He also made a number of suggestions about water supply: of all the great engineers, none was more conscientious than Jessop over the vital matter of ensuring that there was always adequate water. A number of companies were to wish that their engineers had been similarly preoccupied.

The route of the Grand Junction begins by following the natural line of the Brent valley, then turns north to follow the River Colne, taking a line that would have been the obvious choice for any of the first generation of canal engineers. But where a Brindley might have staggered his locks along the length of the waterway, Jessop's greatest concentration, lifting the canal over 50 feet, comes in the six tightly grouped locks of the Hanwell flight. After that comes a steady, if unspectacu-

lar, climb towards the first of the major obstacles, the Chiltern Hills. It seems quite the most obvious route, but originally the plan was to go further to the east with a lock-free section between Hemel Hempstead and Rickmansworth, but with a tunnel at Langleybury, south of Kings Langley.

Jessop was no enthusiast of tunnels – and his experiences on the Grand Junction were to confirm his views. The technology of tunnel construction was borrowed from mining, but it was hedged around with uncertainty. Where a mining engineer could simply find the best way to follow the seam of coal or ore – and if one direction seemed problematical, could set out in an entirely new one – the canal engineer had no choice but to go as straight as possible between the two ends. Surveying techniques were well established. The line was laid out at the surface, and the height above the proposed level of the canal calculated. Shafts were then dug down to that depth, and headings, matching those set at the surface, were worked out from the foot of each shaft. In theory, the different headings met in a continuous straight line; but the kinks to be found in many canal tunnels, particularly the older ones, show that theory and practice did not always coincide. The greatest troubles resulted from ignorance of what lay below the surface. The science of geology was not so much in its infancy as unknown. Indeed, one of its pioneers, William Smith, was about to discover the principles of rock strata through his own work as a canal engineer in the 1790s. Tunnellers set off in the dark, literally and metaphorically. Jessop soon changed his mind about the Langleybury tunnel and decided that lock-building was preferable.

When the canal reached the Chilterns, the solution to the passage through the high ridge was quite different from anything that might have been considered in the early canal age. No locks, no tunnel; instead Jessop called for a deep cutting to slice through the hills. This was civil engineering on a new and impressive scale. It might not look as impressive as, for example, the deep cutting that carries the M40 through the chalk ridge. But one has to remember that the canals were built by muscle-power, not machine-power. Pick and shovel and crude black powder were used, literally, to move mountains; and when the hills had been shifted to carry the canal, the spoil still had to be removed. The quickest way was to carry it straight to the top of the new cutting, but this was impossible for an unaided man with a barrow – so the barrow runs were invented. Planks were laid up the side of the cutting and the barrows attached to a rope. The rope led up, over a pully, to a horse and, as the horse walked away from the edge, the load was pulled up, balanced by the man trying his best to keep his footing on the greasy, vertiginous walkway. Coming down again with the empties was, if anything, even more exciting. The man held the barrow behind him and raced down the steep side of the cutting. It was a job

Hillmorton, Oxford Canal.

A pair of narrow-boats, the motor towing the butty, at Three Locks, Soulbury in the early 1950s, just after nationalization.

that called for some skill and a good deal of raw courage. The same scenes were to be repeated less than half a century later just a few yards away, when Robert Stephenson arrived with his group to create a second Tring cutting to carry the London & Birmingham Railway.

Once clear of Tring, the canal swoops and swings down the hill in a slalom course of locks, which take it past the reservoirs that keep the canal in water and past the Bulbourne maintenance yard. This is a spot where many elements that help to give the canal scene its unique fascination all come together. First, and most important, there is the response of the canal engineer to the environment. At Tring, Jessop met the challenge head-on: cutting brutally through the land in what must originally have produced a pallid, deathly gash of open chalk-face, contrasting with the green of the hills. Time has softened it, but even now the deep cutting is unmistakably a man-made feature, quite at odds with the natural world. Then, as the cutting ends, the engineer has given himself up to the rhythms of the landscape, following the hill contours round, with little jumps down through the procession of locks. The reservoirs, built to feed the Tring summit, are again wholly artificial features which time has blended in, blurring the edges with reeds and providing a home to a great variety of birds. In summer, not only do the birds vastly outnumber their human visitors, but some at least make it clear that – whatever the reason the reservoirs were built – they, the birds, now claim them as their own. Swallows come in like marauding fighters, turning at the very last second to wheel away, when it had seemed they were about to skewer a visitor through the head. The canalside scene shows a typically appealing, yet wholly practical, development. First there are the locks themselves, a broad and stately passage like a gargantuan stairway to some vast baronial hall. Then there is Bulbourne maintenance yard, scarcely changed over the years, still turning out massive lock gates of solid oak; metal is used as well, but the lighter metal gates have proved less satisfactory in practice. They show an irritating disposition to allow themselves to be closed, then as soon as one's back is turned they swing open, moved by a gust of wind or a chance current in the water. The old-style gates close with a satisfactory clunk and stay closed. The yard's only frippery is the little ornate clock tower above the wood-working shop. Add to this the lock cottage, pub and simple bridge, and it would seem that everything has come together by happy accident to create a perfect visual match. But this is no accident, simply a wonderful example of what happens when practical demands are met plainly and simply, using the materials most readily to hand.

A pair of British Waterways boats passing Bulbourne maintenance yard in the 1950s.

Wooden lock gates under construction at Bulbourne yard: they are still made there today.

The canal continues, swinging its way through the countryside until the next major obstacle appears: the River Ouse, crossed by a cast-iron aqueduct. It seems a wholly logical solution to a particular, and not uncommon, problem – getting a canal across a river valley. There are now only hints in the landscape of the troubles that beset engineers for many years. The original idea was to cross the river on the level: locks would lead down to the river on one side and up again on the other. It was soon realized that this would cause water-supply problems in filling the locks, and water problems of another kind, as the Ouse was very prone to flooding. An embankment and aqueduct were the alternative, but by the time this was being considered, in 1804, cash was short and the company was eager to get traffic moving to bring in some revenue. Jessop proposed building temporary locks on the cheap, using timber in place of brick – and he pointed out that the lock gates could be used again elsewhere. The temporary locks could be taking traffic while work went ahead on the aqueduct. A conventional aqueduct was built, but not without a certain amount of grumbling about specifications not being met. In 1807 part of the embankment caved in and the aqueduct was reported to be in a far from satisfactory condition: piers had shrunk, arches were out of alignment. Arguments raged over who was to blame: the company blaming the contractors, the contractors saying their work had long since been approved. The question of who should do the repairs was settled by events. The whole structure collapsed, damming the river and creating fears of widespread floods. There was to be no more talk of repairs. The cast-iron aqueduct was built instead – not the most graceful structure in the world, but at least it has stood now for nearly 200 years. It is a useful reminder that not every plan, even when approved by an eminent engineer, can be guaranteed to work.

Jessop was to face more problems up ahead as the canal headed north towards Blisworth hill. There were to be two tunnels: the first, at Blisworth itself, was to be 3,056 yards long, to be followed by the 2,042-yard-long Braunston tunnel. Both were to give trouble, largely because of the contractors' failure to carry out the work thoroughly – and the failure of the company's own engineering staff to keep them, literally, on the straight and narrow. At Braunston there were, however, unexpected problems with quicksand, which the test borings had somehow missed. But, in spite of these difficulties, the tunnel was duly opened in June 1796. That failure to keep a straight line can still be seen by those boating through the tunnel, in the shape of a watery chicane – an S-bend in the middle. The Blisworth problems were not to be so easily overcome.

Things started to go badly wrong at Blisworth. The contractors began, without Jessop's knowledge, to alter the line, in a direction where the going was easier. The brickwork was shoddy and its inadequacy appeared as sections crumbled and collapsed. The whole tunnel was hopelessly awry, and there was a good deal of discussion over what should be done. Jessop's answer was to abandon the tunnel and opt instead for a deep cutting at the top of the hill, with locks on either side. Steam-engines would pump water back up the flight to keep the summit-level filled.

BOTTOM: *Boats waiting to go through Blisworth tunnel in the 1910s.*

BELOW: *Repair work at Blisworth tunnel, 1910. The picture shows how the tunnel has an inverted arch below the normal water line.*

Beginning at Braunston, the 93-mile canal (now largely known as the Grand Union route to London) runs south via Stoke Bruerne, Cosgrove, Fenny Stratford, Rickmansworth and Uxbridge to the River Thames at Brentford. Built as a wide canal, the Grand Union owes its character to the simple curvilinear arch spans in brick, often limewashed, projecting string courses that conceal the transition between flat-bed coursing and the curved-bed bricks that form the parapets. Hipped slate roofs on pump houses, dry docks and lock cottages - the latter often built into the side of locks - complete the picture.

Barnes, who had conducted the original survey, offered to contract for a new tunnel, which he claimed he could build in three years. The company overruled their chief engineer and took the option of the engineer whose supervision of the first tunnel had not, one would have thought, inspired confidence. Work began in September 1796, but another problem now held it up. The grand optimism that had given birth to the canal mania had died, with the increasing threat posed by war with France, as Britain's European allies began to show their weaknesses. Where once investors had rushed to pour money into canal schemes, now they hung on to what they had. Work on the tunnel came almost to a standstill. But it was still imperative that a through route should be established, and if boats could not be taken through or over the hill, then an alternative had to be found.

The company decided to build a railway – not a railway as we think of it today, but a simple railed track along which trucks could be hauled by horses. The job of building this tramway, as such early railways were known, went to a specialist in the field, Benjamin Outram – who was also partner, with Jessop, in an iron works in Derbyshire. After years of stop-start, go-slow and speed-up, the tunnel was completed and finally opened in 1805. It was, after all, a major achievement: one of the longest tunnels in the country, but wide enough to allow a barge through or two narrow-boats to pass. But the shifting sands and loose rock that had plagued the tunnellers have continued to create problems ever since. Barnes, in arguing the case for the tunnel,

had said, 'Tunnels when completed require little Expense to keep them in repair'. Had the company foreseen just how vast that 'little Expense' was to prove over the years, they might well have opted for Jessop's cut and locks. The main line ended with the junction at Braunston and a little canal settlement grew up there, though its character has inevitably changed over the years. The old boatman's pub now has what appears to be half a stranded Mississippi steamboat tacked on the end. Some things, however, do not change, and one can still admire the quiet elegance of the two cast-iron bridges that arch gracefully across the junction. They also represent an excellent example of how little is needed to bring out their quality: crisp black-and-white paintwork does the job superbly well.

The Grand Junction was, from the first, something more than just a route from the Thames to Braunston. The original Act allowed for a number of branches to be built. Several of these were, in the event, to be discarded, but others were added and altogether the Grand Junction was enlarged by offshoots to Aylesbury, Buckingham, Northampton, Old Stratford, Paddington and Slough. Of these, the Paddington branch was by far the most important. One carrying company, at least, realized its potential even before the branch was completed. Pickfords had already established their own wharf at Braunston and, once they heard of the Paddington scheme, they established another base there. Soon they began a fly-boat service between Manchester and London, running boats day and night, with relays of horses and change crews. They were notable for their efficiency, and the Paddington wharf was served by a regular cart service from central London. Pickfords' success came from their ability to grasp new opportunities. They had begun as local road carriers, then when they saw the improvements offered by canal, they moved on to the water. But come the railways, they were just as quick to leave the canal system as they had been to join it, while the twentieth century saw them back where they started – on the roads.

Of the other arms, the Slough and Aylesbury branches are open but comparatively little used, other than as moorings. The Northampton arm provides a vital link to the delights of the River Nene, while the Buckingham and Old Stratford branches have virtually disappeared. The Wendover arm continues to fulfil its main function of feeding water from the reservoirs up into the Tring summit. There is, however, one other arm, built not by the Grand Junction but by the government: the Weedon ordnance canal.

In 1803 there were real fears of a possible invasion from France, and the government wanted to establish a major arms and ammunition depot in the centre of the country, at a place that offered first-class communications. They opted for Weedon, which could be served both by the Grand Junction and by the main roads now known as the A5 and the A45. The new depot was meant to be more than an arsenal – it was also to be a refuge for king and government if the worst should happen. So the canal led into a walled area with barrack blocks to either side, then continued on to the extensive magazine. The low arch that gave access to the barracks could be closed by a portcullis, a curiously medieval touch to a nineteenth-century canal. The complex was to have a long history of use. In 1809, 22,000 muskets were sent off on canal boats to London, under the guard of two companies of the Bedfordshire Militia. The canal trip was a success, but part of the load was shipped on to a Thames barge at Brentford, which ran aground during a storm. As late as the First World War there was a regular munitions traffic carried on Fellows, Morton and Clayton steamers. But Weedon's days as a military depot came to an end in 1965, and in 1984 it was sold. The complex remains more or less intact – and the portcullis is still down, even if there is no arsenal, and no preparations for a royal retreat.

The appeal of the Weedon arm lies in its curiosity value: even the Royal Military Canal, also built during the Napoleonic Wars to speed the movement of troops to possible south-coast invasion sites, cannot boast a portcullis. But it is, in every sense, untypical of the canal as a whole. The Grand Junction is essentially a workmanlike canal, where frills are kept to an absolute minimum. Where, for example, Sapperton tunnel has magnificent porticoes designed to impress the visitor with the importance of the work that lies behind them, Blisworth tunnel is as plain as plain can be: the canal simply disappears into a dark hole. Plain does not, however, mean dull. The bridges that punctuate the passage of the canal have no frills, but they are so absolutely right, both as landscape features and as part of a practical transport system, that they need none. The design is wholly functional, much as it was on the earlier canals, but on the narrow waterways designed by Brindley the bridges rear up steeply, like caterpillars crawling up a narrow branch. On the broad canal the bridges can ease up the slopes, making their way more gently, so that their shallow curves echo the gentle swelling, the dip and rise, of the land. Ostentation appears only on demand. Where the line passed through Grove and Cassiobury Parks, the Earls of Essex and Clarendon insisted that the canal should be 'ornamental'. And so it is, and the bridge at Grove could well have been designed to grace a landscape by Capability Brown. Topped by a balustrade, with a dentilled cornice, the low arch of dressed stone is a model of good taste. Rather than spoil the perfect symmetry, the towpath passes through a separate arch, balanced by a twin, which is purely decorative, on the opposite bank. Similarly, at Cosgrove, where not only did the main line pass close to Cosgrove Park, but the Old Stratford arm skirted the edge, the simple requirements of engineering had to give way to the demands for architectural embellishment. Here the mock-medieval rules, with what is probably the only canal bridge in Britain with a pointed arch, covered in elaborate decoration. It is all a little absurd, but carried out with such brio and panache that one simply accepts it as a curiosity – and a delightful curiosity at that.

Wooton Wawen aqueduct, Stratford Canal.

It might seem a little odd that such an important waterway should lead all the way from the Thames to a somewhat inconclusive terminus at the, by now, somewhat outdated Oxford Canal. It makes sense, however, when one realizes that the canal was seen from the first as part of a grander design.

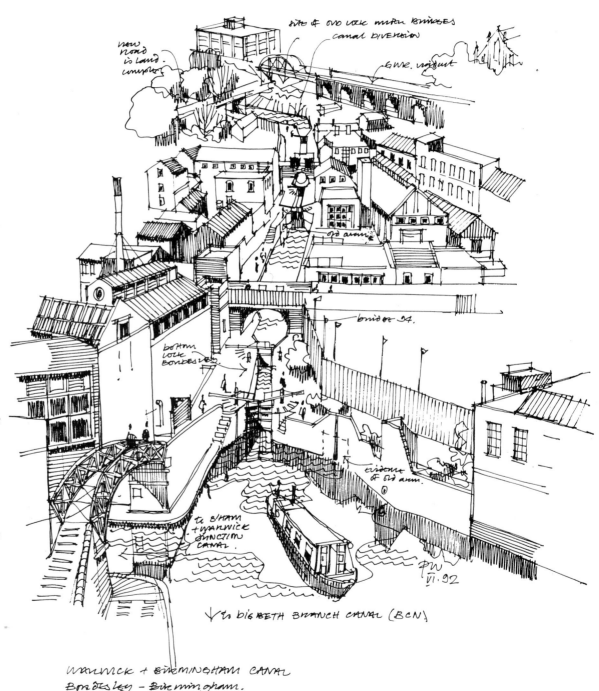

Part of what is now known as the Grand Union, the Warwick & Birmingham Canal starts at Bordesley wharf in Birmingham and heads, via Solihull, Knowle, Kingswood, Hatton and Budbrooke, to its termination at Warwick. At the Warwick end it has been severely truncated, but a waterway project centre now operates at the Saltisford arm, which once led to Warwick wharf and basin, some 22½ miles from Birmingham. At Bordesley there is little of the 1793 fabric left. Locks were rebuilt in the 1860s, and more recently millions of pounds have been spent on canal culverts and road modifications, involving a diversion scheme and replacement locks. The canal to the left is the Birmingham & Warwick junction canal that leads to Saltley.

Among the 1793 Acts was one for the Warwick & Birmingham, the title of which accurately describes its function. It offered a direct route between the two towns, but was still cut off from the Grand Junction. That gap was filled the following year when an Act was passed for a canal from Warwick to Braunston. This was later amended to become the Warwick & Napton, so that a through route was established from the Grand Junction to Birmingham, which was to run over a short section of the Oxford Canal. Inevitably this involved a good deal of legal and financial wrangling. Not only was the Oxford bound to lose trade on its southern section, but there were also problems of water supply. In the event, the Warwick & Napton ended up paying a decidedly handsome toll to the old Oxford company. It was a price worth paying: there was now a modern through route all the way from London to Birmingham. But there was another price to be paid as well – the heavy cost of engineering work.

One thing anyone travelling to Birmingham by canal soon discovers is that the city is built on a plateau, so that from whatever direction it is approached, the boat crew can look forward to a lot of hard work. It all begins easily enough, with the lock-free section of the Oxford Canal, between Braunston and Napton, a thoroughly rural section where the most prominent landmark seems to look back to a slow, easy way of life as the canal skirts Napton hill and its old windmill. After that, the climb begins, gently at first, through the three locks at Calcutt. Beyond that, there are reminders that the canal was built as a working transport route, not to give pleasure to idle dawdlers. Names tell of the old trade – Gibraltar Bridge, the Blue Lias pub – reminders of the busy quarries that supplied cargo to the boats. And all the time, the work rate is being increased – the three Calcutt locks serve as a warm-up for nine more at Stockton. After that, life becomes a little easier again, and travel is more of a pastoral saunter, all the way to Royal Leamington Spa.

The name of Leamington inevitably conjures up images of string trios in the Pump Room, elegant Regency terraces and fashionable shops. This is not the Leamington seen from the canal. When the canal first came this way, the spa had not yet developed, and it in fact developed as one of two very different Leamingtons. The fashionable area kept itself to itself, while along the canal a quite separate town grew up, of industry in the shape of foundries and gas works, and little terraces of workers' cottages. At Warwick, the ancient county town, the canal is banished to the edges, and the arm that once made the connection to the centre has long since been abandoned. As is so often the case, tourist town and canal town have little, if anything, in common. Not that there is much time for the boating fraternity to muse on such things, for up ahead lie the 21 locks of the Hatton flight. What makes them doubly daunting is the fact that much of the flight is in a straight

The Grand Junction Canal served industrial Leamington rather than the fashionable Leamington Spa. This is the Eagle Foundry.

line, so that one stares up at a seemingly endless staircase of black-and-white balance beams. The publican who built his inn at the very top clearly knew his business well. After that, there is a welcome pause, a lock-free section, with only the little Shrewley tunnel as an interruption. At Kingswood there is a junction with another 1793 canal, the Stratford, but the main line continues to head north through what is now Birmingham suburbia, before reaching its conclusion at the Digbeth branch of the Birmingham & Fazeley.

Here not only the Grand Junction and its extension came to an end, but the broad waterway as well. The proprietors of the Birmingham system were adamant: their locks were to remain narrow. It was self-interest, but miserably short-sighted. The Birmingham interests wanted to protect their territory from an invasion by broader, potentially more efficient craft from the south. They were not to know that in little more than a quarter of a century, the land would begin to resound to the hiss of steam, the hoot of whistle and the clank of wheels on metal rails. The narrow-boat had seemed more than adequate to meet the transport needs of the 1760s – it was rather less viable in the 1790s, and it was to seem anachronistically puny in the nineteenth century. It survived, and in a small way still survives, but the decision to retain the narrow locks of Birmingham was a severe handicap to trade throughout the whole system from London to Birmingham. There was little enthusiasm for transshipment from broad barge to narrow-boat, so the system evolved its own answer in time – the working pair. If narrow-boats had to be used, at least time could be saved by fitting them, two at a time, into the broad locks and working them together. But before looking at the trade on the system, we have to backtrack down the Grand Junction to what was to become an important part of the system that built up around the new main line from Birmingham to London.

In 1793 a canal was begun which, from the first, was designed with a link to the Grand Junction in mind. The Leicestershire & Northampton Union was designed to join the Leicester Navigation to the River Nene at Northampton. William Jessop was once again the chief engineer, but as was so often the case, ambition far outran resources. The canal staggered on for some seventeen miles from Leicester and there, in 1797, it rested. In 1799 James Barnes was brought across from the Grand Junction to give his views. He proposed a new connection with the Grand Junction at Braunston, but nothing happened. Three years later he tried again, suggesting yet another new junction, this time at Norton near Crick. The company was dubious and called in another engineer, Thomas Telford, who also proposed going to Norton, but via Market Harborough. The company did not exactly spring into action, but by 1809 it did manage to get the waterway as far as Market Harborough and there, once again, it stuck.

A mile post near Norton Junction.

The Grand Junction Company was understandably irritated by all these delays, and in 1819 it obtained the Grand Union Act for a canal from near Market Harborough to Norton. The Old Union Canal, which started the move out of Leicester, was built to allow vessels of 10-foot beam to travel out from the River Soar; the new canal was to be narrow – a reasonable decision given the immense difficulties to be overcome. The climb up from the Market Harborough line begins straight away at the Foxton staircase: two staircases, each of five locks, with a short passing pound in the middle. To describe this as a bottleneck is a gross understatement. The frustration it must have caused in the working days of the canal can now only be imagined. Various schemes were proposed – widening the flight, duplicating the locks – but only one came to fruition, the famous Foxton inclined plane, opened in 1900.

The inclined plane was not a new idea. It had first been introduced into Ireland in the 1770s and made its way to England in 1788 on the private Ketley Canal, which served the Shropshire iron works – an area famous for innovations of many kinds during the eighteenth century. The iron master, William Reynolds, was the man largely responsible, and there were three of these remarkable railways for boats in the course of a mere seven and a half miles. One of them, at Hay, has been restored as part of the Blists Hill open-air museum site. The inclined plane has already been briefly described, (p. 42) but it is worth elaborating a little here to

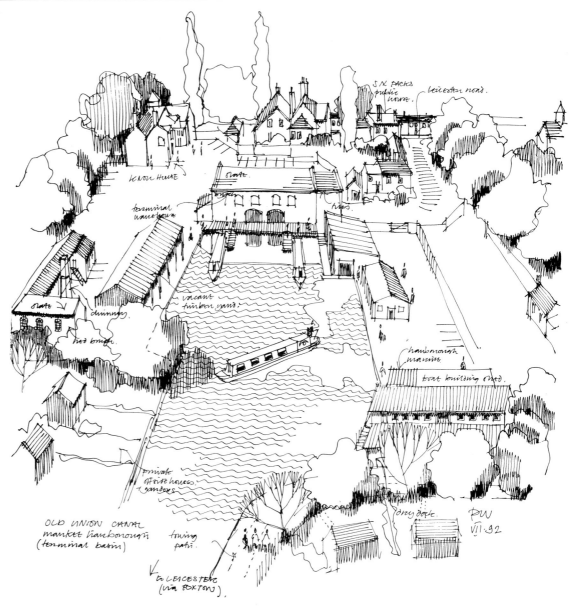

As built, the Old Union Canal extended from Market Harborough wharf to West Bridge, Leicester. It was bought by the Grand Junction Company and was later to form part of the Grand Union Canal. The waterway was eventually extended south, via the locks at Foxton, but originally terminated at the basin, a mile or so outside Market Harborough, on the Leicester road. The brick terminal warehouse, adjoining timber yard and boat-building concern are part of a redevelopment project, designed in due course to restore the fine, listed buildings and to increase the visitor appeal of this interesting market town.

make it clear how great an advance the Foxton plane was over its predecessors. For a boat to go downhill, it is first let into a lock with a wheeled carriage set on the bottom. As the water is let out, so the boat subsides until it is resting on the platform. The carriage is designed with small wheels at the uphill end and large ones at the downhill, so that when it runs out on to the railed track that leads down the hill, the platform remains horizontal. The descent is controlled by cable, and the weight of the loaded boat going downhill is used to raise an empty vessel on the next track. The boats were simple tub-boats – exactly what their name suggests, for they look not unlike slightly overgrown bath-tubs.

It was not the length of the Foxton plane that distinguished it from its forerunners, for the lift was only one-third of that at Hobbacott. At Foxton, however, it was not little tub-boats that were to be moved, but full-scale narrow-boats. Here, instead of a simple cradle, caissons were used – huge tubs of water into which the narrow-boats were floated, two to a caisson. Once again, the two caissons were balanced. It was not quite the pioneering project that it is often portrayed as being. An incline able to take barges up to 70 feet by 13 feet 4 inches had been built on the Monkland Canal in Scotland to complement, rather than replace, the flight of locks. It was completed in 1850 but had already been closed for a decade before work on Foxton began. But Foxton itself was destined for a far shorter life. In 1910 this 'wonder of the waterways' was a sorry sight of rusty rails just visible in a sea of weeds. Today it has been cleared again, and at least one can stand at the end of the slope and imagine what an extraordinary sight it must have been, as the boats bobbing about in their watery containers crept slowly up and down the banks.

Long before the innovations came to Foxton, however, there had been great changes at either end of the main-line route. At the London end it was clear that while the Paddington basin was doing good business, the centre of trade had moved eastward to the new docks, such as St Katharine's, which were being developed along the Thames. In 1812 work began on the Regent's Canal, which was to continue the line on down to Limehouse, where a new dock was to be constructed, joined to the Thames by a lock able to take ships up to 350 feet long.

Over the years Limehouse Basin – or the Regent's Canal Dock – as it was originally and more grandly known – was to turn into a mêlée of craft of all sizes. Sailing ships, barges and lighters jostled for space, while the little narrow-boats hustled and bustled between them. If ever there was a clear message that the scale of narrow-boat operation was too modest for growing needs, then the message could be read here. It was reinforced as the wide barges made their way out through East London, en route to Paddington or further afield. This was also a canal that was perceived in different ways. In 1885 the skipper of a Thames sailing barge, which was making its way up the canal, vented his feelings at the frequent groundings in round, though no doubt censored, terms:

They never tells yer w'en they're goin' to lower the water, nor nothin'. It's never clean, an' its allers low water, and there's nothin' but naked men a bathin' and thieves wot robs your barge and takes all they can git out of 'er, and blackguard boys wot calls yer names and spits on yer, and throws stones at yer.

The canal was a busy trading highway, 'greedy for stone and timber and coal and hay and all the material that a great city demands from the country'. But it was, from

the first, thought of as a pleasant amenity. In the original plans it was to pass right through the middle of Regent's Park, which would (the promoters assured the world) be greatly enlivened by the colourful passage of boats. The world, however, decided that the only things likely to be enlivened were the ears of fashionable company, as the language of the 'bargees' floated up from the water. The canal was banished to the edges of the park. Now that trade has virtually ended, the amenity has come into its own. Little Venice really does have something of the romantic charm that the name suggests, and the pleasure-boats throng where once the laden barges made their ponderous way. Yet for all its charm, there is an air of sadness that lingers over the empty wharves and quietly crumbling, begrimed warehouses. The stalls of knick-knackery that crowd around Camden Lock make for a bright and colourful scene, but the canal was designed for more solid fare. Its old function has died – speeded, in part, by the development (or rather lack of development) at the other end of the line.

Canals began to spread out through the Birmingham area, and in 1784 a long-running dispute between the Birmingham Canal and the Birmingham & Fazeley was finally resolved when the two companies combined to form what was to become Birmingham Canal Navigations or, more simply, the BCN. By the beginning of the nineteenth century, the BCN consisted of a complex web of some 70 miles of canal spread out across the Birmingham plateau, but at its heart lay the old winding, twisting Birmingham Canal laid down by Brindley. Major improvements came when Thomas Telford was called in to create a new main line, literally cutting through the middle of the old, for it is the deep cuttings that give the line its character. The best known of these lies in Smethwick, where it is crossed by the delicate arch of Galton bridge. Its name is altogether appropriate, for it is named after Samuel Galton of the Lunar Society, a body of distinguished men who met to discuss the latest scientific and technological advances of the day. Galton bridge leaps the deep cutting in a single 150-foot iron span. Something of its drama has diminished, since as part of a road improvement scheme, the canal was pushed down into a new tunnel, in effect little more than an overgrown concrete drainpipe. By way of compensation for this indignity, the twentieth century has supplied one new and dramatic bridge, carrying the M5 not only across the new main line but over a waterways 'flyover' for the old Brindley canal had to be relocated in an aqueduct. Telford, one feels, would have approved.

The old Brindley canal did not wither and die when the new, improved version appeared. Too many industries had taken sites along the banks for that to occur, so the old route now became a series of loops, which modern pleasure-boaters negotiate at their peril. I remember, all too clearly, the occasion when my propeller engaged with some indescribable mass of wire and cloth, which defied all attempts at clearance by the conventional method of immersing one's arms in the blue-green greasy

Balance beam step, Stoke Bruerne locks, Grand Union.

liquid that serves for canal water and hacking away with pliers. On this occasion the boat had to be manhandled into Bradley yard, where the stern was hoisted clear to reveal the horror in all its glory – at which point it was all too clear why puny pliers had proved ineffective. In the event an oxy-acetylene torch was needed. That, however, is the loop today, a backwater of decayed memories of former industrial grandeur. A century and a half ago such a loop was still very much part of the busy trading scene. The BCN had become an amazingly complex system of highways and byways, but for all the improvements one aspect had not changed – the locks remained narrow, and any boat that wanted to trade up from London at the heart of this system needed to be narrow as well. The days when the canals could proudly proclaim their role as the leading transport route of the age were coming to an end.

In the 1830s the navvies who had laboriously dug the deep cuttings of the improved Birmingham Canal found themselves being offered work by a new engineer. Thomas Telford was now an old man, and a young, vigorous Robert Stephenson was in charge of work on the railway from London to Birmingham. It is a tribute to the surveying skills of Jessop that Stephenson's line should so closely follow his for

Ted Barrett Senior at the tiller of the Samuel Barlow butty, John.

so much of the route, but it was a compliment the canal proprietors would have been happy to forego. All down the line the new railway with its hissing, speeding engines offered all too visible a challenge to the canal and its horse-drawn boats. Change was coming at a frightening rate: it seemed that almost overnight the crude, ponderous machines that had chuffed along like portly, overweight gentlemen were being replaced by sleek new machines, which were travelling at speeds that only a few years ago had been ridiculed as impossible and unattainable. The narrow-boat load – so much better still than anything the roads could offer – seemed suddenly a good deal less impressive. Competition had unmistakably, and frighteningly, arrived.

The impact on trade was not immediately apparent if one looked just at the volume of goods being carried: in fact, in the years immediately following the opening of the railway, trade on the Grand Junction showed a modest increase. But the value of that trade told a very different story: from around £150,000 a year in 1837 it dropped to just over £80,000 ten years later. The first to suffer were the fly-boats, which had offered a fast service based on relays of horses and change crews – but canal boats could not compete with railways, no matter how fast the horses galloped. But the canal-side factories and works still wanted their steady supply of coal and raw materials, which could be delivered right up to the loading bay by canal. Even then, however, prices had to be slashed and the pattern of trading began to change.

In the new competitive age, some companies left the canals altogether. Pickfords had made its profit in fly-boating; now it turned to the railways. The small company and the individual boatman could no longer live on the small profit-margins, and trade moved inexorably towards the bigger companies. The Grand Junction was soon running its own fleet, which had grown by the 1850s to be the biggest in the country. Rationalizing the system was all very well as far as the larger companies were concerned, but the changes forced on the boating community were altogether more dramatic. In the old fly-boat system, with its changes of crews, the job of the boatman was not unlike that of the modern long-distance lorry driver. The men worked in shifts during the dashes up and down the country, and returned to their families for a break before the next run began. The move to the slower, steadier trade meant less furious dashing around the countryside, and where four men had once been employed to keep a boat running day and night, now only a couple were needed. And, all the time, prices for cargoes were dropping – and price reductions inevitably found their way through to lower wages. It became more and more difficult for one boat to support two men and their land-based families; and correspondingly sensible for a family to take on a boat for themselves. And if that family was going to spend most of its time on the move, then who needed a house that would stand idle for most of the year? The boating family was not a novelty in the railway age, but it was railway competition that forced the pace and turned what had been an exception into more or less standard practice – so much so that when one looks back at the working life of the canals, the image that comes to mind is of a working family crammed into the tiny confines of the back cabin of a narrow-boat.

Throughout the rest of the working life of the canal system based on the great trunk route of the old Grand Junction, the narrow-boat reigned. Improvement was difficult. An obvious answer was to turn, as the railways had done, to the transforming power of steam. Yet once again, the limiting factor was the size of the locks, which meant that there was simply no way in which the size of the boat could be increased. Put in a steam-engine and you used up cargo space – worse than that, still more space was needed for the coke to keep the engine running. Fortunately, at least as far as the Grand Junction was concerned, the steam-engine could be made sufficiently powerful to work two boats; unfortunately, it required a great deal of fuel to do so. As a result, the steamers became the natural successors to the fly-boats, moving swiftly and silently up the canal on a regular time-table. The slow, irregular trade that made up the bulk of canal travel still depended on the horse well into the twentieth century.

The true age of power boating had to wait until the second decade of the twentieth century. There was a brief flirtation with the gas-engine – in reality, not much of an improvement over the steam-engine. It was the diesel and the semi-diesel engine that brought the breakthrough, largely due to the work of a Swedish engineer, Erik August Bolinder. His new engine was given its first water-borne trials aboard a Thames lighter in 1910, and by 1912 the Grand Union carrying company, Fellows, Morton, had installed one in the narrow-boat *Linda*. The company had another vessel, *Lynx*, being fitted out for steam, but one trial was enough – steam was abandoned in favour of diesel. The Bolinder was a somewhat idiosyncratic machine. Before it would work, the fuel had to be vaporized in a hot bulb by the somewhat alarming expedient of heating it with a blow-torch. Surprisingly, perhaps, the use of blow-torches in engine rooms dripping hot fuel produced no major catastrophes. Once safely under way, the heat of combustion kept the engine running with the characteristic popping of the two-stroke engine. Boats, however, are required to stop as well as go forwards, and the Bolinder had no reverse gear. The trick was to slow the engine down as far as possible, and then pull a lever for a quick injection of fuel that caused a violent backfire, which brought everything to a shuddering halt. This was all very well, but the boatman had to be quite sure that his vessel really was revving very slowly, or the backfire failed and the boat careered cheerfully on, running down another vessel or charging up a bank. Life was seldom dull with a Bolinder.

Lock gate collar, Kennet & Avon.

Even after the establishment of the Bolinder and other petrol and diesel engines, there were always those who preferred the old ways, working with horse or mule. This was particularly true of the Number Ones, the boating families who owned and worked their own boats. For them every ton of cargo carried made an addition to a low income – and every bit of cargo space sacrificed to motor and fuel was money lost. For these people, life on the canals went on in the middle of the twentieth century much as it had done in the middle of the nineteenth. It was a life of unremitting hard work, starting as early as four or five in the morning, when the first task of the day was to water, feed and harness the horse – and not ending until perhaps eight or nine, when once again the horse was the first to receive attention. The animal's needs always had priority – on the horse's strength and endurance all else depended. A fit shire horse could pull a pair of boats holding as much as 60 tons of coal all day and every day at a steady one to two miles an hour and, given a pair of empties for the return journey, it could positively gallop along, covering up to 50 miles a day.

There were no frills in a Number One's life, and profits could disappear a lot more easily than they could be earned. There were all the regular expenses of boat maintenance and care of the horse, plus the ever-present threat of calamity. A horse might

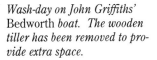

Wash-day on John Griffiths' Bedworth *boat. The wooden tiller has been removed to provide extra space.*

die, a harsh winter might see boats immobilized in the ice for weeks on end, trade could falter or stop altogether. The wonder is that anyone should choose such a hard life, with so little material reward. Most, of course, did not choose it: they were born to it. The boat families, for ever on the move, their boats their homes, had, as the years went by, less and less connection with families 'on the land'. It became one of those systems that feeds on itself: the further away the boat families moved from the rest of society, the more the mutual distrust grew. Children, on the rare occasions when they had the chance to get into a conventional education system, found themselves unwelcomed by their fellow-pupils and treated with scorn by all too many of their teachers. As a result, they learned little and only the most adventurous among them were prepared to push their claims to learn more. The boating community became ever more inward-looking, not because it necessarily wanted to cut itself off from the rest of society, but rather because society wanted to keep it at arm's length. This was not by any means unique: an old coal miner told of how his mother was determined to keep him out of the pit and managed to get him a job in a local factory. But before he had been there a week, the foreman came up and said, 'Your dad's a miner.' The lad agreed that he was. 'Then go and join him.' And he was handed his cards. The boat family, like the mining family, formed a closed community. Strangers were by no means always welcomed, and strangers very rarely welcomed them.

This might give the impression that the boating families had a lonely, miserable life, tied to the routine of moving cargo endlessly around the system. This was far from being the case. Many boats worked regular runs, up and down the same length of canal; many a family working on the Grand Union would never see anything of, for example, the canals north of Birmingham. The Grand Union was their home, just as a town or village might be home to a family on the land. It had its local pubs, where they could be sure of meeting friends – one of the more appealing aspects for many holiday-makers is the frequency with which pubs appear alongside the canals. Many of the pubs grew up with the canal system, and this was reflected in their names, the Boat, the Ship, the Anchor or, quite simply, the New, to distinguish it from the old established pubs and inns of the area. At places where boats regularly gathered, such as junctions, the pubs inevitably became boating locals – pubs such as the Greyhound at Hawkesbury, or Sutton Stop, as it was known to the boating community. Others developed their special associations, because landlords went out of their way to make boat people feel at home. A good example was the Bridge Inn at Etruria on the Trent & Mersey. When Josiah Wedgwood established his great pot-works here in the eighteenth century he built a large estate of cottages and small houses for the workforce. In the nineteenth century a boatman bought one of these cottages and used it to sell beer to his old

mates off the cut. The enterprise was such a success that the cottage next door was bought, the two knocked together and the Bridge Inn was born. It remained in the same family, scarcely changed by the years, until it was demolished to make way, inevitably, for a road-widening scheme.

Boating remains thirsty work, even for those of us who do it for fun, and the canal pub continues to fulfil much the same function as it always did: a place to tie up for the night, to relax, to exchange stories of the day's doings. But there have been changes. The last thing the working boatman wanted at the end of the day was reminders of work. The pub was a good, plain, honest building – indistinguishable from any town pub in a working-class district. The holiday boater wants to feel the old pub's connection with the past, so canal memorabilia abound in the shape of old photographs, painted Buckby cans and the like. When the old warehouse at the foot of the Audlem locks on the Shropshire Union was converted into the 'Shroppie Fly' pub in the 1970s, the bar was created out of part of a narrow-boat. Other new pubs took up canal themes in their architecture – not always, it has to be said, with any great success. The Jolly Tar at Barbridge Junction, where the Middlewich branch leaves the main line of the Shropshire Union, has indeed a jolly nautical theme, but the array of portholes somehow suggests that it should be heading out for mid-Atlantic, not down the line to Birmingham. The desire of the holiday-maker to feel part of the old life of the canals is altogether understandable – the mistake is to believe that one is actually experiencing anything even approximating to that life. It was the people that gave the old canal pub its character. Take them away and it could, to modern eyes, seem a bare, rather cheerless place. Many a canalside pub has managed to preserve the balance between the good, honest boozer and the no less honest modern pub, enlivened with old prints and memories of working days. Some have gone too far, while others have simply abandoned the waterways' past altogether, and one or two actively discourage anyone from the boats using the place at all. The changing role of the canal pub reflects the changing role of the canal itself – it has altered, just as the canal has altered. The memories of working days survive, but they are just that – memories. The same is true of the canal as a whole.

Wharves, warehouses and stables once proliferated along the canal, as did buildings put there specifically to take advantage of the improved water transport. Today, it is not always easy to imagine, let alone see, what the system was like in the last century. Rickmansworth is a case in point. That this was once an important, if small, canal settlement is clear from the little cluster of buildings alongside Batchworth lock. Offices and stabling of warm red brick have recently been converted, quite successfully, and include a small canal centre, with an old Ovaltine boat outside for local colour, for Ovaltine ran its own fleet of boats for many years

Crinan Basin.

on the Grand Union, carrying coal to the works. But this only hints at the waterways activity that went on in this area. Still visible is the lock that gives access to the River Chess, which was canalized to give access to a wharf in the town centre and was mainly used by Salters brewery. A neat canal house stands in the little strip of land between the two waterways. Less obvious is the fact that there was also a second branch opening opposite the Chess, serving Batchworth mill, and another by the bridge to Bury Grounds, and a baker who took delivery of flour by water. The town had its very own narrow-boats, universally known as 'Rickies', and built by local boat-builders, Walkers of Rickmansworth. To the canal passer-by, on foot or by boat, the canal now seems a marginal affair, skirting the edge of the town, a thing with little impact or influence on development. Investigation shows just how important the canal was in reality. Today a little of its old importance has returned – not as a transport route, but as a picturesque landscape feature. Trinity Court has been built alongside in fashionable neo-vernacular, or Lego-styled building: the car park is gabled, and a tower with an incongruous pyramidal top suggests that a church has been mated with a warehouse. It makes the un-selfconscious finesse of the older canal buildings seem all the more appealing.

Other reminders of the working days are less immediately obvious. The traditional placing of bollards at broad locks was not devised in order that working boats could remain steady during the rise and fall of the water, but to help with the stopping of boats entering the lock, so that no time was wasted. Everything had to be done at top speed, and the method of working a lock would bring howls of outrage were any modern pleasure-cruiser to try it. Paddles were drawn as boats came in, so that the flow of water helped close the gates – when travelling up hill, the motor would back up and charge the gate to force it open. All kinds of methods were used to speed the boats on their way, and the dexterity with which ropes were whipped round bollards and gate posts could have earned many a boatman a job in a Wild West circus. Today, anyone working up a long flight of locks, such as those at Hatton, can be quite certain that the boatman of 50 years ago would have got up in half the time. One cannot imagine – no, that is not perhaps quite true, one can all too easily imagine – the comments of a boatman faced by modern hydraulic gear, which not only has to be wound slowly up, but has to be wound equally slowly down. It is all a matter of emphasis: the holiday-maker is assumed to want an easy life; the working boatman wanted the fastest possible journey.

There is an assumption now that everything should be done to make life as easy as possible for the holiday boaters, but is the assumption justified? Everything one sees as pub decoration, souvenirs, even holiday boat decoration, harks back to the old working days. It may be merely a nostalgic whim to believe that this should be extended to the canal itself and its structures. There is a sense of completeness about a Grand Union lock, which is not just a visual matter – sound too plays its part. The satisfying thunk of wooden gate meeting wooden gate, the gurgle and rush of water as paddles are drawn, are gentle, soft noises, perfectly matched with the harsh rattle and clang of a ratchet paddle gear. Even the visual side is important in the long term, for as one moves around the country, so one finds that the different engineers all had their own ideas about the mechanism to be used. The Grand Union might have its own consistency, but its mechanisms were very different from those of neighbouring waterways. Much of the appeal of canal travel lies in its diversity. It is not just a case of a meandering Brindley canal contrasting with a more straightforward route by Jessop; the fascination lies in good measure in seeing how different men produced different solutions to what was, in essence, the same problem. It might be a major problem, such as how to advance the canal when a large hill lay inconveniently in the way, or it might be a minor problem, such as how to get water in and out of a lock chamber. It is worth remembering that what we know as the Grand Union system has been in existence for two centuries, and over that period of time working boatmen, engineers and officials refined its workings to make it as efficient as possible. That is not the same as saying it was as easy as it could be: to use the system as well as humanly possible requires skills that few, if any, weekend boaters – or canal authors – possess. But at least, with a traditional waterway, one has a glimpse of the unattainable ideal.

The narrow-boat ruled the Grand Union and its branches. One can travel down to Brentford and see the barges and wide boats that still appear near the Thames and ponder what might have been, but in practice the narrow locks of the Birmingham heartland set the pattern. Jessop and his colleagues created a system that was as efficient as any that could have been contrived in the conditions of two centuries ago. Over the years, generations of boating families learned to use that system just as well as it could be used. Now, when the narrow pleasure-boat has taken over from the working narrow-boat, the system still offers a challenge – to try to use it well. None of us will, in all probability, take up the challenge and be content with the result; but there is a deep satisfaction to be had from coming within a reasonable distance of the ideal. When the boat slides smoothly in, the gates close as the stern clears their path, and the paddles are on the move with their congratulatory rattle a blink later, then one can tie up for the night where the old boats tied up, pop into the pub that has heard tall stories of boating exploits for some 200 years, and enjoy a well-earned pint.

THE SHIP CANALS

The narrow canals of the Midlands and the wide ship canals are the two extremes of the canal age – yet the notion of the ship canal has its origins in a time long before narrow canals were dreamed of. Since medieval times, the great rivers had created inland ports, usable by some of the biggest ships of the day: not that they were necessarily that large. The caravels, the favourite vessels of so many of the great explorers who set out to discover new worlds, were seldom much more than 70 feet long by 25 feet beam. This compares, roughly speaking, with a Thames sailing barge, the main difference being the deep keel needed for the ocean-going caravel. Apart from the problems faced by boat owners caused by lack of water depth in the natural rivers, there was also the difficulty caused by other river users. The river was not just a means of transport, it was also a source of power – turning water-wheels to work a great variety of machinery from the familiar grindstones of the corn mill to the bellows of the blast-furnace.

The boat owner and the mill owner had one interest in common: they both wanted to control the waters. The boats had no wish to shoot rapids or come aground in shallows; the mill owners wanted an assured, regular and steady supply of water. For both interests the solution was the same: construct a weir across the river, so that it fell in a series of controlled steps and did not plunge around in a totally haphazard manner. The miller could now tap off his supply from the deep water close to the weir, by means of an artificial channel or leat. The boater could negotiate this watery step by a system of removable paddles in the weir, which allowed him to ride down on the released flash of water, or be winched up against the flood. The trouble was that when the boatman wanted his release, the miller found his leat running dry. There were endless quarrels and arguments, and in some cases landowners solved the quarrel in their own way – they built the weir on a fixed structure, with no flash lock, no moveable paddles. They closed the river to

A ketch-rigged Severn trow at Gloucester, with Albert Mills in the background.

navigation; and if they were powerful enough they got away with it. This is what happened at Exeter, and it gave birth to Britain's first ship canal.

In the twelfth century the Earls of Devon built a weir across the River Exe, and the citizens of what had until then been a prosperous trading city found their route to the sea barred. No amount of pleading and petitioning had any effect on generations of earls, and it seemed no power could be raised to force them to do anything against what they saw as their own best interests. But however powerful an earl might have seemed to even the wealthiest citizen, there were others who outranked them. In 1539 Hugh Courtenay managed the none too difficult task of falling out with Henry VIII, who promptly had him arraigned for treason and his head duly fell. This was just the chance for which Exeter had been waiting, and this time their petition for a channel that would restore the city as a port succeeded. Exeter was now at the height of its power, with a prosperity firmly based on the booming wool trade. The country, as a whole, however, was an economic shambles. A war with France, into which England blundered as an ally of Spain, was a military disaster and an even greater financial one, resulting in a debased coinage and inflation that ran totally out of control. There were rebellions all over the country, with some of the fiercest opposition coming in the West Country, where Exeter found itself under siege. There was no time, money or enthusiasm for canal-building: the promoters of a later age may have thought they had problems seeing their scheme through to completion, but their difficulties were nothing compared with those faced by their forebears. It was 1566 before the canal was built that would bypass the Countess Wear (*sic*), which had kept ships out of Exeter for four centuries.

The Exeter Ship Canal has a special place in the history of Britain's waterways, for it was the first to introduce the pound lock, the familiar lock closed off with gates at either end. The credit for devising the lock belongs to Italy, where an anonymous engineer built a canal with 18 locks in 1458, but these had gates that rose and fell like a portcullis, and the superstructure got in the way of shipping. It was that great polymath Leonardo da Vinci who introduced the mitre gates of the type generally in use today. So when John Trew subsequently introduced them to Exeter, he was about a century behind continental practice. Nevertheless, the canal was an impressive venture, if not quite the canal we know today. It was scarcely what we would call a ship canal – a mere 16 feet wide, with a depth of just three feet – but the locks were grand affairs. They were designed to take a group of vessels at once, and the two lower ones were 189 feet long and 213 feet wide. Although they were pound locks, they were unusual by today's standards in being built with turf sides – a style of building still to be seen on the Kennet Navigation section of the Kennet & Avon Canal. The canal too was less impressive than it is today. Just under

Towing down the Exeter Canal.

two miles long, it ended at Topsham. Today the canal is five miles long, the locks have been improved and it will take genuine, if modest, ships – up to 122 feet long, 26 feet beam as far as Topsham lock and, more importantly, up to 11 feet 6 inches in draught. The impressive docks and warehouses at Exeter date back only as far as an improvement scheme of 1824. Their rugged simplicity, the rough sandstone blocks contrasting with massive doors, which look as though they could adequately have kept the revolutionaries out of the Bastille, are all very different from the urbanity of the 300-year-old Custom House.

The notion of improving rivers by adding artificial cuttings, or extending the navigation further inland than the natural waterway would allow, was to be pursued with enthusiasm both before and during the canal age. Two of the schemes authorized at the height of the mania in 1793 were of just such a kind. The Foss Navigation was designed to fulfil two quite different functions: to provide a route for trade into the countryside north of York, and to drain the land. The first survey was carried out by the ubiquitous William Jessop in 1791. It was never a major waterway. For the first few miles it followed the River Foss, then cut across a bend of the river above Strensall, in a canal that ended at Sherriff Hutton. No one seemed very convinced by this dual-purpose land drain and waterway which, once it left York, made its way through the Vale of York and, it seemed, stopped when the rising

ground brought it to a halt, rather than because it had reached any particular or important destination. It probably needed the wild enthusiasm of the mania years to bring it into being – and it required another Act in 1801 to raise more money to complete the work. Considering that this was a simple navigation, running for a mere 12½ miles through largely flat land, the second Act could be taken as a resounding vote of no confidence in the project on the part of local investors. Those who did put their cash in were disappointed. By 1859 all but the first mile and a quarter was abandoned; surprisingly what is left has a certain grandeur.

Today, the Foss is navigable from the Ouse just below Skelder Gate Bridge and boasts within its short length one lock and one basin. Memories of former glories appear in the shape of immense castellated red-brick warehouses. The Foss, though, has been reduced to little more than an approach to a handy basin for city trading. For once, the eighteenth century had done no more than restore a river to something like its former importance, providing a trading route for the City of York just as the Vikings did for the settlement of Jorvik.

OPPOSITE: *Laggan locks, Caledonian Canal.*

BELOW: *Considered as a natural defence by the Romans, the River Foss was made navigable to connect with the River Ouse at Blue Bridge, downstream from York itself. Now reduced to under two miles, the navigation passes the former Yearsley Basin and the York Union Workhouse. The castellated Leetrams Mills have been converted into offices and flats, and new apartments overlook the pondweeded Castle Mills basin locks, and the Castle Museum and Law Courts beyond Tower Street.*

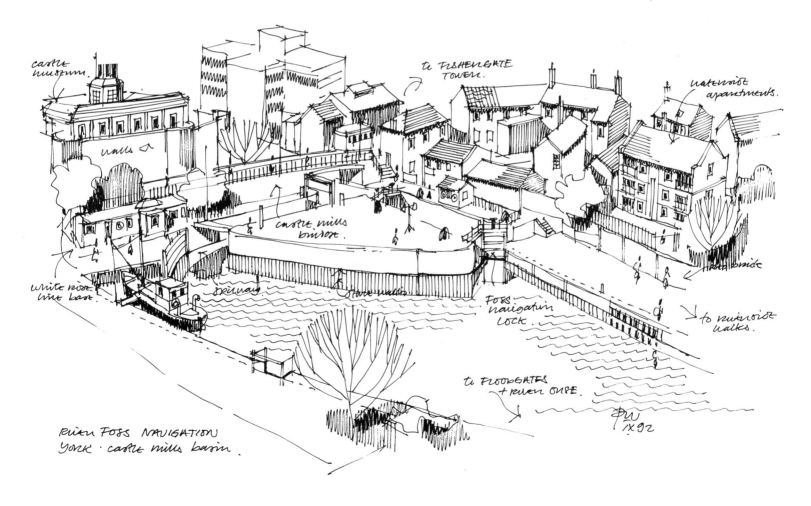

The second navigation scheme was a good deal longer, a good deal more successful, and a good deal more curious. The idea of joining Chelmsford to the coast was explored as early as the 1760s, when John Smeaton proposed a scheme for a 13-mile canal, costed at £16,697, but nothing came of it. In 1793 a new scheme for making the Chelmer navigable was brought forward – the engineer was John Rennie. It was very much a river navigation, with 13 locks and a series of short artificial cuttings. At the Blackwater end, however, it seems very much of ship canal dimensions. This is the heart of Thames sailing-barge country, and the lock allows vessels up to 107 feet long and 26 feet beam to make their way off the tideway into the safety of Heybridge Basin. It is when the navigation heads inland to Chelmsford that the system changes its character. It is not just that the locks shrink to allow vessels 60 feet by 16 feet to pass, but the water levels drop even more dramatically. The maximum draught falls to a paltry two feet. The result was that the special barges, although they looked quite grand, broad-beamed vessels, could carry a cargo of only 25 tons. It seems extraordinary that so much effort should have gone into building a broad waterway, with 13 locks, and yet to build it to such a meagre depth. It reflects the description of the waterway in Priestley's review of the country's navigations and canals, published in 1831: 'The chief object of this navigation is the supply of Chelmsford and the interior of Essex, with coal, deals, timber and groceries, and for the export of corn.' This does not sound like a recipe for prosperous trade, nor was it. Cynics were known to remark that the only real success of the waterway came from the willows grown, and harvested, along its banks. The Chelmer & Blackwater, and the Foss, really highlight the fact that the great age of improved river navigation was over – old schemes could still be made better, but new schemes, based on minor rivers, were likely to remain, literally, backwaters. If there were to be new, major advances, then something altogether more drastic and dramatic was called for. Two other new routes of 1793 were to provide just that.

The River Severn was already one of the great trading rivers of Britain when written records began. As early as 1430 an Act of Parliament laid down that 'The River Severn is common to all the King's liege people to carry and recarry all manner of merchandise as well in trowes and boats, as in floats otherwise called drags.' The Act is interesting not just for its affirmation of the central importance of the river as a transport route, but also for its mention of 'trowes', or trows, as they are more generally spelt. Over the centuries, the trow developed as a versatile sailing barge, which evolved a characteristic shape to its hull, with a transom stern, rounded bilges and slab sides. The bottom was flat. Originally they were square-rigged, but later they became fore-and-aft-rigged and by the nineteenth century they were quite capable of trading round the coast. The vessels were sturdy and their skippers needed to be highly skilled to cope with the swift currents and dramatically high rise and fall of the tide in the Bristol Channel.

The trows were at one time able to trade upriver as high as Shrewsbury, lowering their masts to pass under bridges along the way. Even larger vessels worked the tideway, but as vessels became bigger, so they found it increasingly difficult to negotiate the shallows and sandbanks – not to mention the occasional appearance of the Severn bore – on the journey between Berkeley Pill (a point on the river bank near Berkeley) and the city of Gloucester. The bold answer was to bypass the section of river with a broad canal, able to take the largest vessels of the day, on what became the Gloucester & Berkeley Ship Canal. It would also make a direct connection with the Stroudwater Canal at Saul, and thereby to the Thames & Severn, and so on to the River Thames itself. This was to be canal-building on a grand scale – a canal initially designed to take vessels up to 15 feet draught, but extended before completion to 18 feet. If the ambitions were great, the actual achievement did not measure up.

The work was at first entrusted to Robert Mylne, a highly competent man whose experience came more from water supply than water transport, for he was the engineer to the New River that supplied London. Like most chief engineers, he was responsible for overall planning but left the everyday management to the resident engineer, Dennis Edson. Given Edson's record, this was perhaps not overly wise, as he had been dismissed from two previous canal workings. He was not to fare much better here: he was gone in nine months, to be replaced by the 26-year-old James Dadford. Work went on painfully slowly, and the proprietors began to take a dim view of, as one of them put it, the notion, 'that an Engineer may render us sufficient service, by thinking and contriving for us, while sitting at his ease in London'. The work was not going well, although various attempts were made to introduce new and more efficient machinery, including a device for removing spoil. Powered by steam, the machine worked rather like a conveyer belt and was said to be able to remove 1,400 barrowloads of earth in a 12-hour period. By 1799 the canal had been dug only from Gloucester to Hardwicke, a less than inspiring five miles away. Cash had run out and attempts to re-start the project failed. Jessop was called in to advise, but that only brought vituperation rather than action.

It is a work that requires other talent & knowledge than a common Canal Cutter. From the time of Jessop's visit, I date its misfortunes. He and Dadford are mere drudges in that confined school; and both are without any sense of extended honour.

Damning criticism indeed – until one discovers that the words were written by the former, and now sacked, chief engineer, Robert Mylne.

The Gloucester & Berkeley languished, and it was the government that came to its rescue. The Poor Employment Act of 1817 authorized Parliament to lend money for schemes that would provide work for the unemployed, and as a result the Canal

Company was allowed £60,000. By now it had been agreed that the other terminus of the canal should be at Sharpness Point, and work was soon under way. A new engineer was in charge, John Upton, and the final section of work went to the London contractor, Hugh McIntosh, for a sum calculated with nice exactness at £111,493 15s. 11d. It was in 1827, more than 30 years after work had begun, that the canal was finally opened. There was not much cash left for ceremonial – two boats and a band were the best the impoverished company could manage. It was perhaps the final appropriate comment on the whole affair that three local youths tried to liven up the proceedings by firing their own salute from a cannon, which promptly blew itself to smithereens.

The fight to get the canal open had been a long one, but the proprietors could at least take pride in the fact that it was the grandest in the country. Over the years there were to be further improvements. The basin at Sharpness proved to be inadequate, as vessels queued and jostled for place while they tried to get out of the river and into the canal. In 1871 work began on a new dock complex. It was all on a very different scale from the work that began back in the 1790s. Portable steam pumps mounted on a railed track kept water out of the dock; stone brought down the canal was unloaded by

Construction of the new dock wall at Sharpness in 1874. The metal sheeting is supported by timbers.

Dock workers at Sharpness. In the foreground is the horse used for shunting on the dock railway system; behind the men are the pipes that sucked up grain from the ships' holds into the silos.

steam cranes; clay dug out was made into bricks on the site; and the area echoed to the repeated rumble and crash of blastings, as holes were prepared for the foundations of a new pier. It was all a sign of the canal's success – and there were further indications at the Gloucester end, where new warehouses were added in the middle of the century. So successful was the canal that, far from seeing the railways as a threat, it welcomed them and actively promoted the building of a railway bridge across the Severn to link Sharpness to the coalfields of South Wales and the Forest of Dean.

A century ago the Gloucester & Berkeley epitomized just what a thriving waterway should be. Even in the Depression years that were to come in the 1930s trade remained brisk, with fleets of narrow-boats, run by the Severn and Canal Carrying Company, feeding goods into the ship canal system. The Worcester & Birmingham Canal had been bought up, so the boats could come down from Birmingham, then continue on down the Severn, often towed by tugs. There could be as many as 20 boats in a single tow, arranged in two parallel rows of ten. This must have made life extremely interesting for the steerers of the last pair, who had little idea what was going on, with the tug perhaps a quarter of a mile away ahead. Now the scene has changed; boats and ships no longer crowd the docks, but many of the old buildings and structures have survived, even if they have had to find a new role. A trip along what is now the Gloucester & Sharpness Canal is full of interest.

Stand at Sharpness at low tide and you can see why the canal was needed. Sandbanks and mud flats emerge, round which the river whips in brown, foam-flecked streams. Today, Sharpness shows more signs of former prosperity than it does of present trade, but it still has an air of grandeur, of being built on a majestic scale. Even the row of houses for the dock workers has nothing mean about it – an avenue of suburban villas, widely spaced, that might well have slipped away from some Laburnum Grove or Acacia Avenue to enjoy the river view. From the river bank, the piers reach out into deep water, opening like a funnel mouth for the ships to pour in. Looking at the vast tidal basin with its huge lock gates, one would love to be transported back in time to the opening day in November 1874 – even though it was cold, windy and pouring with rain. Out of the gloom came the first customers. Tugs hauled in two Italian steamers loaded with grain, to be followed by the magnificent square-rigger *Protector*, which had sailed across from Norway, and the barque *Director*, stacked with timber from New Brunswick. Operating the lock gates was itself a fascinating job in those early days: first the giant capstans were turned to allow water in or out, then the gates were swung by a busy, bustling little steam locomotive. But hydraulic power – more efficient, but less exciting – soon took over.

Arranged around the wharf is a strange assortment of buildings. A tall, concrete grain silo looks down on the red brick of the old Severn Ports Warehousing Company. The former, being in effect little more than a giant storage bin, is feature-less, while its neighbour is full of patterns and contrasts. Loading doors reach from ground to eaves, and the space between is filled with rows of shuttered windows – brick, stone, wood and iron – the rhythms and contrast are those of so many canal-side buildings, warehouses and mills, here presented on a grand scale. Modern warehouses line the newer quays, but the old familiar features have changed little over the years. Bollards and rings do the job they have always done; railway lines snake in and around the dock, poking in towards every shed and silo. But as you move down towards the old dock, dereliction starts to appear. Boards bang in the old sheds, gaps are filled with limp, grey plastic; sand blows in and piles up in corners. Beyond these lies the original entrance lock, where rusty lighters are tied to rusty bollards, and the only vessels are the bright, new sailing boats designed for pleasure, not commerce. The railway link that once brought trade across from Wales has gone. On a foggy night in 1959 two tankers collided, and hit the bridge in a disaster that covered the whole river in blazing oil. Two whole spans were carried away, and the bridge was never repaired; instead the arches that remained were shipped off to Chile, where they now support a road. By the canal is a circular stone pier, which once supported the section of the viaduct that could be swung to allow ships to use the canal.

The swing bridge is very much a feature of this waterway. The expense of building bridges high enough to allow tall-masted ships to pass underneath would have been immense. The bridges come complete with little houses for the bridge-keepers.

These are somewhat curious affairs, not unlike lodges at the approaches to some grand country house. Looking, for example, at the house at Splatt Bridge, near Frampton-on-Severn, what one sees from the canal seems to consist entirely of a portico. A little door lurks bashfully behind an array of fluted columns. The building seems impossibly small, until one looks at it from the back and discovers that the house is actually built against the bank, so that what at first seemed like a single-storey lobby is actually a two-storey cottage. The motif is carried round, with pediments on a more modest scale on the other three sides. These cottages seem universally referred to as 'handsome' and 'Regency' – terms that are generally thought of as interchangeable – but, in truth, they are all but overwhelmed by their ponderous fronts. Regency architecture is noted for its balance; here it over-balances.

The bridges also provide focal points along what can otherwise seem a somewhat featureless canal. The waterway runs broad and straight, raised up above the fields. Church towers stand like beacons proclaiming the presence of village or hamlet, but in general the ship canal pays little attention to such minor distractions. Frampton-on-Severn might boast the biggest village green in England, but that was

Constructed to avoid the dangerous reaches of the River Severn, the ship canal links Gloucester, via Hempstead, Saul and Purton, to Sharpness docks, a distance of some 16 1/2 miles. Canal swing bridges are operated both mechanically and automatically, and at several sites the bridge-keeper's cottages are still extant. They feature the 'Greek revival' style, with charming fluted Doric porticoes, and are stucco-rendered – conveying an elegance and form not out of place in either Cheltenham or Royal Leamington Spa. Although now worked automatically, Purton upper bridge still has its keeper's cottage, with a low-pitched slate roof. Nearby is the church and, at Purton lower bridge, the waterside Berkeley Hunt pub.

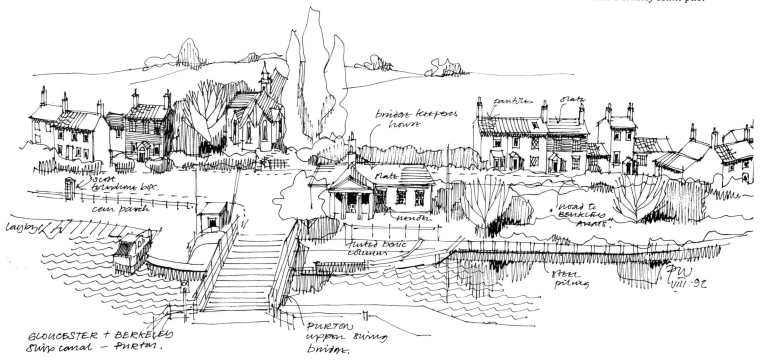

no reason for the owner of a cargo steamer to call in. But if villages were of little interest to canal traders, the surrounding country was, and grain warehouses still dot the route. Beside Fretherne Bridge is the old Cadbury chocolate factory, with its wharf and covered loading bay with a long canopy. At Sandfield Bridge there is what appears at first glance to be an oversized canal junk shop, with bits of swing bridges, cogs and winches scattered around. Beyond is Saul Junction, where a surviving mile-post indicates that the journey is half over, with G8 on one side and S8 on the other. Here is a working boatyard and the link with the Stroudwater Canal. To the west are the ruined remains of a lock, but to the east the canal is in water and is home to a long line of boats.

As the canal approaches Gloucester, so the banks become ever more crowded with concerns such as timber yards, built to take advantage of the cheap water transport. The grand finale arrives at Gloucester itself.

To put Gloucester into perspective it is worth looking at the street still known as The Quay, and still announcing its old function by a line of bollards at the water's edge. Until the coming of the canal this was, in effect, the port of Gloucester. Set against that, the new dock area with its immense warehouses looks doubly impressive. The first dock basin was completed in 1810 and the inner basin in 1848, and over the years new warehouses and buildings were added. The town got a rail connection right at the start with the construction of the Gloucester & Cheltenham tramway, built to bring stone from the quarries at Leckhampton down into the dock. The oldest warehouse is now the Pillar Warehouse of 1836, which gets its name from the pillars that carry the upper storeys out over the wharf. The remaining warehouses are very similar to the older warehouse at Sharpness, and there is a reminder that this is a true inland port in the Mariners' Chapel, dwarfed by its towering neighbours.

Gloucester docks like Sharpness made extensive use of rail connections. It was not the railways that killed off trade. In the 1920s there was a regular trade of steamers on the canal. Vessels of less than 300 tons were allowed to make their own way, 'provided they employ a licensed Canal Pilot and Hobblers'. The latter used ropes to help control the ship in the narrow waters. Larger vessels were taken down the canal by tug. It was road transport, not rail, that brought carrying days to an end, but the docks themselves have found a new life, at least partly connected with their old traditions. The National Waterways Museum has found a home in and around the old Llanthony Wharf warehouse. As at Ellesmere Port, there is as much interest in the vessels outside on the water as there is in the more conventional museum exhibits inside. And they are not just boats moored up and inert – the old steam dredger is regularly set to huff and clank in the dock; vessels are

Skew bridge, Macclesfield Canal.

The old river quay at Gloucester, with the new canal dock behind it. The little lock cottage is dwarfed by the tall warehouses.

moved and exhibits from other museums made welcome visitors. The museum is new and has already proved to be immensely popular and very successful. It also provides a superb example of how an old building can find a new use and still preserve its integrity. It is not many years since this area was a scene of desolation, with buildings growing sadder and more dilapidated with each passing year. It seemed that their days were coming to an end. Ironically, the motorway that now carries the trade that once came by water now boasts huge new signs enticing the motorists off, to visit the 'Historic Docks'.

The Gloucester & Berkeley was not the only ship canal to be authorized in 1793, though it was by far the most ambitious. The Ulverston Canal was a mere mile and half long, designed to bring ships into the town to a new basin. John Rennie was given the job, and by 1796 the canal was open, and within a few years was doing immense trade. Iron and copper ore from surrounding mines and slate went out, while coal and timber came in. By 1846, on average, over 500 ships a year were using the canal – but then the Furness Railway opened and numbers began to decline. In 1862 the Railway Company bought the canal and docks at a knock-down price and by 1945 they were closed; and sealed off from the sea. The canal that once took coasters is now used only by fishermen, and the remains of the old sea lock are reminders only of what the canal once was.

The last of the 1793 ship canals has enjoyed an altogether happier fate, though it never reached that state of glorious profitability so confidently predicted by its promoters. The objective was simple: to save ships the long sea route round the Mull of Kintyre by cutting across the narrow neck of land between Lochgilphead and Crinan. The nine miles of canal would replace over 130 miles of sea passage. The only problem – one says the only problem, but it was in fact a huge one – lay in the hills that rise up along the spine of the peninsula. The canal required ship locks at either end to provide access, but in between them are 13 locks to lift the waterway up to the summit and back down the other side. An all too familiar story soon began to unfold. The money ran out, more was applied for, but even then savings had to be made. The canal begun to very reasonable dimensions, able to take craft 88 feet long and 20 feet beam with a draught of 15 feet, was reduced at the western end, and the depth reduced to a meagre ten feet. Even vessels designed with use of the canal in mind have been known to find this water-level something less than adequate. The Clyde

Linking the town of Ulverston to the entrance sea-lock into Morecambe Bay – a distance of one and a half miles – the Ulverston Canal commences at the foot of Hoad Hill, about quarter of a mile outside the town. Overlooked by the Folly lighthouse, the basin end features roadside wharf buildings in both stone and rough-cast render under stone slate roofs, together with the Canal Tavern public house. Once accessible by 100-foot-long vessels with a draught of 12 feet or so, this little ship canal has a road along its northern bank, which apparently was never used for towing purposes. At one time 120-ton coastal vessels and steamers were able to effect a passage, but today the sea-lock is derelict and silted up.

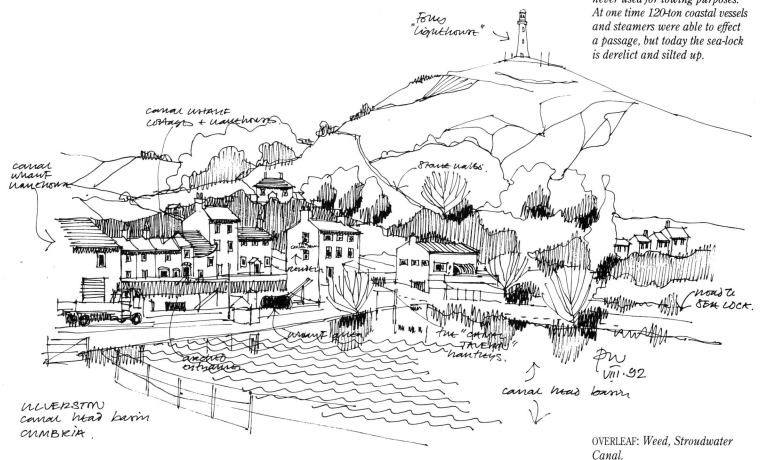

OVERLEAF: *Weed, Stroudwater Canal.*

puffer once traded throughout the waters of the west coast of Scotland. Steam-powered, flat-bottomed to run up on beaches, it was the all-purpose workhorse of the highlands and islands. One of the last survivors, the VIC 32, built for the Admiralty during the Second World War, now plies a new trade, carrying holiday-makers, but every year it makes the canal passage from its home port of Ardrishaig. Regulations give the maximum speed on the canal as 6 mph – an unobtainable dream, it seems. Two mph seems good going, as the iron hull bumps and grinds along the bed of the canal and the vessel moves forward at the speed of a Zimmer frame, though the canal still finds favour with a variety of yachts and the occasional fishing boat. Those who do travel the canal find it a pure delight.

At Crinan, the basin sits snug in a little ring of hills and cliffs at the edge of Loch Crinan. Then the canal simply follows the edge of the estuary, with its broad, gleaming flats, and twists its way round to follow the contours of the river. As the road from Lochgilphead keeps to a very similar route, drivers are occasionally presented by the somewhat alarming sight of a sizeable coastal cruiser coming round the corner towards them. The canal continues curving round the rocky outcrops of a heather-covered hillside until it begins the climb up to the summit level. Then

Pleasure-steamer and fishing boats at Ardrishaig at the eastern end of the Crinan Canal.

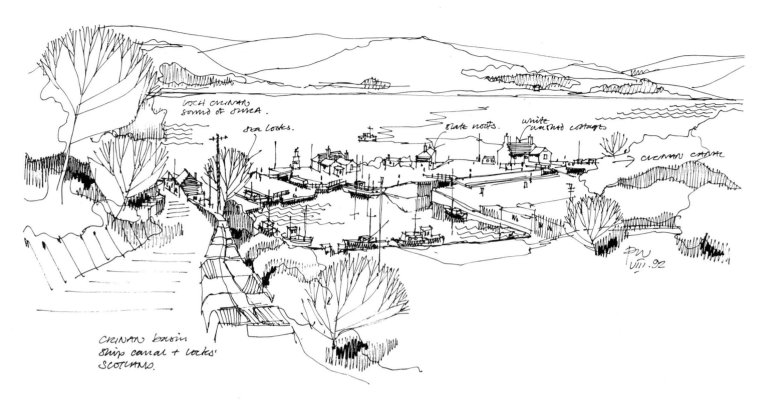

there is a brief respite, before it is time to start coming down again towards the long sea-water Loch Fyne, famous for its Highland scenery and even more so for its delectable local kippers. Ardrishaig itself still has a workmanlike air, with a large timber wharf beside the sea loch.

The relative success of the Gloucester & Sharpness as opposed to the Crinan Canal comes from their very different functions. The Crinan was never more than a short-cut to save a journey that comparatively few vessels undertook in the first place; the Gloucester served a busy inland port and had connections to a rapidly growing and prosperous industrial area. The lessons would have seemed to be moderately clear, but a few years later a new scheme, of even grander dimensions, was brought forward for a canal to save another long, and often perilous, sea route. In 1802 the Caledonian Canal was approved, to cut across not a narrow peninsula, but the whole of the Scottish mainland, from Loch Linnhe near Fort William in the west to Inverness in the east. It was not so much one canal as a series of canals linking together a chain of locks, including Loch Ness, which (to those who have ventured on to its waters) can seem like an inland sea. The notion was not simply that of building a canal because a canal was needed. The Highlands of Scotland were in an appallingly depressed condition – due in no small measure to the activities of the

Extending from Ardrishaig to Crinan, this ship canal obviates the passage around the Mull of Kintyre. Whitewashed cottages with fixed ladders on chimneys (to facilitate sweeping), together with roller swing bridges, contribute to the navigation's character. The scenery is splendidly Highland, with forests and reservoirs to feed the canal summit. As a storm shelter from Loch Crinan and the Sound of Jura, the basin at Crinan contains fishing boats and yachts and enjoys the occasional passage of a 'puffer'. Built as an employment programme, the canal is a reminder of splendid Georgian enterprise and philanthropy.

landlords, who took the land from the crofters for sheep-farming or to create sporting estates. The Highlanders had suffered the misfortune of fighting on the losing side in the Jacobite risings and now they were paying the penalty. The government sent Telford to report on their condition and to suggest public works that would bring employment: one result was the Caledonian Canal.

The Caledonian was, without question, the most ambitious scheme of the canal age. Telford was to be the engineer, but much of the credit for planning and designing the works rested with his mentor from the Ellesmere Canal, William Jessop. Among those who came to see the work in progress was the poet Robert Southey, who wrote in detail about what he saw and rhapsodized about the majesty of the scene, where man vied with nature. At the western end, the canal passed beneath the great slumped shoulder of Ben Nevis, Britain's highest mountain – and the canal responded with its greatest engineering feature, the eight-lock staircase at Banavie. 'The pyramids,' declared Southey, 'would appear insignificant in such a situation, for in them we would perceive only a vain attempt to vie with greater things.' Here, however, he said – very much in the manner of a man of the Industrial Revolution – was a work of art that was, above all else, useful. Yet oddly enough, the staircase does not impress visually in the way that, for example, the Foxton and Bingley staircases do. There they rear up steeply and dramatically; here the locks themselves are so large that the effect is minimized, for the top of the flight is a quarter of a mile away from the bottom. The staircase at Fort Augustus actually seems the more impressive, simply because the vertical rise is so much greater.

It is one of the features of the Caledonian Canal that its immense works of engineering suffer vividly in comparison with far lesser works on other canals. Ask a canal enthusiast to name the canal with the most impressive deep cutting, and the chances are the answer will come back – the Shropshire Union. Yet the Laggan Cut is a far greater work in every sense: broader and deeper, and no one could complain of lack of drama in the surrounding scenery. This is partly the problem: the Shropshire Union cuts create their own drama; everything around them is lost from view. On the Caledonian one is always aware of the scenery of mountains and hills. More importantly, however, it is a question of scale: the canal itself is so broad that there is none of that claustrophobic closing in of the sides that makes deep cuttings in the narrow canals so memorable.

Perhaps the most remarkable feature of all is in many ways the least obviously important. At the eastern end the canal meets the shallow bay of the Beauly Firth. Even at high tide, large vessels cannot get close to the shore, so if vessels could not reach the canal, then the canal would have to go out to meet the vessels. Clay was brought from a nearby hill on a little railway specially built for the purpose, and was

Split bridge, Stratford Canal.

gradually built up into a bank that poked out into the sea. It extended for nearly a quarter of a mile from the high-water mark, and when it was finished, stones were piled on top and it was left to consolidate. Six months later work was re-started. A channel from the canal was dug into the top of the mound and a chamber for the sea lock dug at the far end, while a chain-and-bucket pump and a steam pump kept the workings dry. Finally the lock gates were hung and ships could enter. Southey was impressed all over again, and wrote a long poem in praise of Telford and his works. It can still be seen, inscribed on the wall of the headquarter buildings at Clachnaharry, extolling all the engineer's work from the Pont Cysyllte aqueduct to the road bridge across the Menai Straits.

The Caledonian Canal was not quite the enterprise it should have been. It opened as a makeshift affair in 1822, able to take fishing boats but not the ocean-going vessels for which it was built. Money was, inevitably, at the root of many of the problems. The canal finally reached its intended size in 1847 – but by then the railway age had arrived and the need for cargo vessels to travel either round or through northern Scotland had vanished. It was destined never to make much more in revenue than was needed to keep it in repair, and there was never the slightest chance that the cost of construction would ever be regained. There were even arguments as to whether it provided any employment at all for the impoverished Highlanders. By 1812 it seemed

Benavie on the Caledonian Canal. The poles sticking up at either side of the lock were used to work the capstans that controlled the sluices and gates.

*Fishing boats lined up along
the Caledonian Canal at
Clachnaharry.*

there were more workers arriving from the Glasgow area than from the mountains of
the west. The records do, however, show one gloriously optimistic view of what might
be done to bring the Scots navvies to a respectable way of life. A small brewery was
established at Corpach 'that the workmen may be induced to relinquish the perni-
cious habit of drinking Whiskey'.

The Caledonian was the one work of the canal age that saw the introduction of
mechanization on any large scale. Southey wrote with wonder of the dredger whose
chimney 'poured forth volumes of black smoke' and of the mechanical digger
'where the temperature was higher than that of a hot house, and where machinery
was moving up and down with tremendous force, some of it in boiling water'.
Impressive as it all was, in comparison with anything else to be seen during construction on the canals at the time, it all seems quite puny compared with the forces
brought together at the end of the century for the last major work of its kind in
Britain, the Manchester Ship Canal.

Almost a century had gone by since the heady days of canal mania, and in the meantime an even more frenetic railway mania had enveloped Britain and subsided again. The idea of a ship canal to Manchester had first been mooted in the 1820s, but nothing had come of it. Various plans were aired in the 1870s, but the real birth came in 1882 when a local enthusiast, Daniel Adamson, called a meeting of locals and dignitaries, together with a London engineer, Hamilton Fulton, and the engineer of the Bridgewater Navigation Company, Edward Leader Williams. The Canal received enthusiastic support from Manchester and Salford Councils – and attracted vehement opposition from others, notably the local railway companies. Transport routes between the Mersey and Manchester seem to have been unusually prone to long and expensive Parliamentary battles, which usually left the lawyers as the main beneficiaries. The Duke of Bridgewater had struggled to get his canal approved against the determined opposition of the old river navigation owners. The Bridgewater Canal Trustees had, in their turn, been equally adamant that the construction of the Liverpool & Manchester Railway heralded a period of doom, the like of which had not been seen since the plagues of Egypt. Now the wheel turned again, as the fight went on for the new canal. Three Bills were needed, hundreds of thousands of pounds were spent, but in 1885 the Act was finally approved.

After a shaky start, when it seemed for a time as if public support for the canal was not going to be translated into cash to build it, work got under way in 1887. Civil engineering had made immense strides since the last ship canal had been built across Scotland. The railway companies had built on the grand scale, and the contractor for the main works, T.A. Walker, had just left one of the greatest of them all, the railway tunnel under the Severn, which was opened in August 1886. Walker was to die before the Manchester was completed, and the whole works were flooded out in 1890. But construction went on, and vast resources were employed. Human manpower was very much in evidence, as it had been on the earlier canals – over 16,000 men and boys were employed in the works, but they no longer depended as they once had on pickaxe, spade and barrow. Great excavators mounted on trains deposited spoil straight into the waiting trucks. Where Southey had been impressed by the few steam-engines set to work on the Caledonian, he would have been astonished at what he found here. There were a hundred steam-excavators, including 55 Ruston and Proctor steam-navvies and two French and three German land-dredgers. And 173 locomotives hauled 6,300 trucks on over 200 miles of temporary rails. Add to that literally hundreds of steam-pumps, steam-cranes and portable steam-engines, and it is easy to see why, at the peak of the works, the canal builders were burning 10,000 tons of coal a month.

Bellanoch Post Office, Crinan Canal.

A tug with a train of barges, dwarfed by steamers on the Manchester Ship Canal.

Everything was on a gargantuan scale. The biggest of the entrance locks had gates each of which contained 230 tons of timber and 20 tons of iron, which had to be opened and closed with hydraulic machinery. As on the Gloucester & Berkeley Ship Canal, road bridges had to be made to swing clear to allow ships to pass: once again the principle was the same, but the scale very different. The bridges were made of steel, with wrought-iron plates for the decking and, when it was built, the swing bridge at Salford was the biggest of its kind in the world. It was not just roads that had to be diverted. Two railways had to be realigned and built up on high embankments so that they could cross the canal, and the Manchester Ship Canal also had to cope with the pioneer of 1760, the old Bridgewater Canal. The stone aqueduct over the Irwell had to go, to be replaced by a new wonder – the Barton Swing Aqueduct. It is a remarkable structure, a steel tank 235 feet long, 18 feet wide and six feet deep that moves on a central pivot pier, its 1,400-ton weight borne by 64 rollers. When it has to be moved, simple gates close off the two ends, and the great water-filled tank begins its stately movement – and never seems to leak or splash. Brindley would surely have been pleased that when his pioneering aqueduct finally had to be pulled down, its replacement was such a magnificent piece of engineering.

The Gloucester & Berkeley and the Manchester Ship Canal, though separated by nearly a century, are essentially similar. Both were designed to revitalize an inland port – and both were to be overtaken by events. Manchester, the grander scheme, was to survive longer as a working waterway, but it too succumbed to changes in scale, as container ships overtook the old cargo steamers and trade moved inexorably from the north of England to the south. On both canals, traffic was soon limited to the seaward end. The M6, the late twentieth century's answer to transport needs, scorns the delays of swing bridges. It rises gently to cross three waterways – the Mersey, the ship canal and the Bridgewater – before subsiding equally gracefully on the other side. As the motorist crawls across behind the slow-moving trucks, there is ample time to see the broad canal stretching out on either side, and not a ship in sight. If it no longer has a role as a major transport route, what is to be done with it? The same question arose for virtually the whole canal system, and for much of that system an answer was found.

Boy workers employed on building the Manchester Ship Canal, c.1890.

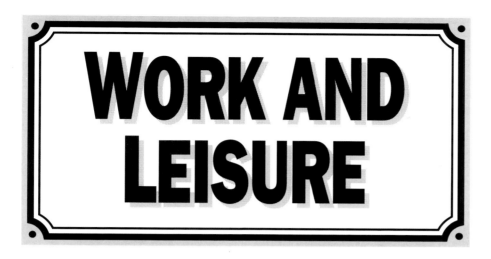

WORK AND LEISURE

There are no records to show just how many people became canal enthusiasts through a first introduction to the Llangollen Canal, but the number could well run into many thousands. The canal seems to offer so much to the pleasure-boater. It is a wholly rural canal, passing through delightful countryside, but it has its moments of – literally – high drama, with a grand aqueduct at Chirk, followed by the even grander crossing of the River Dee on the Pont Cysyllte. Many holiday-boaters are happy to leave it at that, accepting the thrills and delights without ever looking back at the history of this canal. They would, perhaps, be surprised to learn that it was not known originally as the Llangollen canal at all, but as the Ellesmere, and far from being thought of as a peaceful backwater, it was promoted as a major route, absolutely guaranteed to return handsome – and with luck exorbitant – profits to those who could grab shares in the enterprise. The Ellesmere Canal was born at the height of the mania years, and the rush for shares was impressive, even by the standards of those days. At the first promotional meeting some 1,500 speculators pledged almost a million pounds in cash to the scheme. The *Chester Courant* reported the event in September 1792:

> Shrewsbury, about 16 miles from Ellesmere, was so crowded on the nights before and after the meeting that many people found very great difficulty in getting accommodated: several gentlemen being obliged to take care of their own horses, cook their own victuals, and sleep two or three in a bed; and so difficult was it to procure horses and carriages from Leicester and Market Harborough . . . that six gentlemen from the latter place actually hired and went in a mourning coach.

Now this is an intriguing item. What on earth were the inhabitants of Leicester and Market Harborough doing, dashing over to Shrewsbury to buy shares in the Ellesmere Canal? The most likely answer is that they were speculators, moving from canal promotion to canal promotion, for the Old Union and the Ellesmere were con-

Early scenes of pleasure-boating on the Ellesmere Canal near Llangollen.

temporaries. Even so, there must have been reasons for chasing Ellesmere shares – and in the 1790s those reasons would not have included its potential profits as part of the leisure industry. So what was the appeal?

The obvious starting place for any investigation is Ellesmere itself, yet the modern visitor will not find any very obvious features that would justify raising £400,000 to pay for a canal to the little market town. It was, it is true, at the heart of a prosperous agricultural district, famous for cheese-making. A new works for turning out rennet had just been established there, but rennet was not exactly the ideal commodity for canal transport. The real answer was to be found further west, where coal mines and iron works proliferated – notably in the area around Ruabon. In fact, the canal could more reasonably have been named the Ruabon, though it was never to reach there: it stopped short a little way beyond Pont Cysyllte, and the final part of the route was to be continued by a tramway. Llangollen was something of an afterthought. Water supply, that perennial problem that beset all canal engineers, could be solved by diverting water from the River Dee. The canal, however, was crossing the Dee and heading off towards Ruabon on an aqueduct that stood 120 feet above the river. Pumping would have been exorbitantly expensive, so the only answer was to tap the river at a point above the level of the canal, and that was some six miles upstream. Here a weir was built – now known more romantically as the Horseshoe Falls – and the water diverted along a narrow feeder. This was made navigable (if only just), so that boats could trade up as high as Llangollen. The section of canal that seems today to epitomize the beauty and tranquil pleasures of the area was, in fact, no more than a minor branch.

There were to be many notions over what was to be the best route for this canal, many revisions of plans and changes of direction, but the one central, pivotal point remained sure and certain – this was not intended as a rural canal, built for pleasure; it was not aimed at Llangollen, from which it now takes its name. At its heart was the busy industrial complex of Ruabon. So how did it come about that this very industrial canal should now be so admired for reasons that have nothing whatsoever to do with industry? The best way of answering that question is to look at the canal itself, both in terms of what we can see today and in relation to its development.

The starting point might seem a little bizarre, for it takes us back to the early years of the canals – back indeed to the work of James Brindley. In 1772, an Act was passed to:

> enable the inhabitants of the ancient Sea Port of Chester and the inhabitants of the interior parts of the country to enjoy more fully the advantages of the Sea Port and of the navigation of the River Dee from the City of Chester to the Sea.

Chester was in grave danger of losing its role as one of the premier ports of the north-west – due in large measure, it has to be said, to the reluctance of the city fathers to spend money on such essentials as the improvement of the docks and the dredging of the Dee. Now work on the new Trent & Mersey Canal was posing another threat to an already declining trade. The original idea had been to build the canal to Middlewich, to give Chester access to Trent & Mersey trade, but in the event it never got any further than Nantwich Basin – which is not even very close to Nantwich.

It was in its way a majestic canal, offering broad locks to take barges from the River Dee, but it remained an isolated affair, divorced from the rest of the system with no very obvious source of trade. Oddly enough, though, it did manage to give a hint of things to come, for it was carrying holiday-makers as early as 1776. They were not, however, interested in the joys of boating: they were travelling from Beeston to Chester for a day at the races. By the time the Ellesmere Canal proposal was being discussed, it must have been obvious that the Dee had lost out to the Mersey and that the chance of reviving Chester's maritime role had slipped away. If anything was to be done with the Chester Canal, then it would have to be extended.

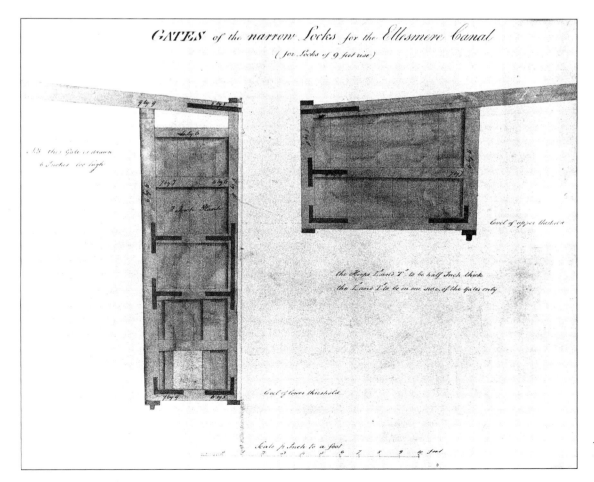

Original designs for lock gates for the Ellesmere Canal – short gates at the top, long gates at the lower end.

Various notions were put forward, which by the early 1790s had roughly settled down to an extension from the Dee to the Mersey, a line from Chester to Shrewsbury and a section through Ellesmere to the coal mines and iron works around Ruabon. The present system emerged out of a welter of proposals: the major routes through the Wrexham area and down to Shrewsbury were abandoned altogether, but a number of branches were added, notably from Welsh Frankton to Llanymynech to connect with the proposed Montgomery Canal. At the Mersey end, a whole new dock system was created and named after the little market town where it all began – Ellesmere Port.

Few holiday-boaters now make the journey to Ellesmere Port, but those who do get a genuine flavour of the working canal, partly through its environment and partly through the Boat Museum that has been established there. On a good day, one can look out across the Mersey to the towers of the Liver Building, which proclaim the importance and prosperity of the Port of Liverpool. It is not difficult to see why the Mersey might offer a more appealing route for shipping than the Dee, with its narrow, twisting approaches through shoals and sandbanks. The new port prospered, helped a century later by the arrival of the Manchester Ship Canal, opened in 1894. The modern works and refineries grew up around the ship canal, and the modest Ellesmere Canal scarcely registers as a waterway among its grander neighbours. It was not always so: over the years a huge complex of docks and warehouses developed, with, at their heart, the 'winged warehouse' designed by Thomas Telford and completed in 1843. For years, this warehouse stood as a superb example of how a building could be constructed to answer purely practical needs, yet at the same time be as visually satisfying as a Gothic cathedral. The main warehouse building had the familiar pattern of loading bays reaching from ground to eaves, with the walls in between punctuated by windows, setting up a satisfactory visual rhythm. What distinguished the warehouse were the wings, great arches that strode across the broad waterway, under which narrow-boats and barges could slide to be loaded and unloaded in all weathers. Above the arches themselves were a further two storeys of warehouse space. Sadly, as so often happened, when the warehouses' useful days were ended, they were saved from official destruction but there was no active preservation. As a result vandals wandered in, and in 1970 Telford's building was reduced to a smouldering wreck.

Ellesmere Port is by no means all doom and gloom. Many old buildings survive, and many old installations. The most exotic is the pump house, where steam-engines were installed to deliver water under pressure to work the hydraulic lifts, cranes and machines of the dock complex. But the principal attraction is the boat collection: wide boats, short boats, narrow-boats and tugs – and their rich variety helps explain the complex system of locks of different dimensions that served the

Hire fleet, Concoform Marine, Grand Union.

Ellesmere Port in its heyday. One of the famous 'wing' warehouses, built on arches over the water, can be seen in the background.

different basins. In Chapter 3 there was a brief mention of the Number Ones – the boating families who worked their own boats. Among the last were the Skinners, and after Joe Skinner's death, their old boat *Friendship* was put into retirement at Hawkesbury. His widow, Mrs Rose Skinner, had a home on land, but most days she would go down to her old floating home and stand in the back-cabin doorway, watching the world go by and talking to old friends and new acquaintances who passed the time of day with her. Now Rose Skinner, too, is dead – but *Friendship*, looking far more spruce than she did in her latter days, has a place of honour in the museum. It is a reminder that many of these vessels were not simply cargo-carrying boats; for many they were both home and workplace, the most important things in their owners' lives. At Ellesmere Port there is a real sense of the working world of the canals, in all its rich diversity. Everywhere there are examples of what the photographer Eric de Maré called 'accidental sculpture': bollards shaped by years of use until they have acquired something of the smooth, gentle sinuosity of a sculpture by Moore or Hepworth; the flash of colour from painted boats seen through the lattice of a bridge; and the ever-present, ever-changing reflections of boats on water. Yet however delightful the patterns, there is nothing contrived – everything relates back to the world of cargoes and working boats. It is not perhaps the world that the pleasure-boaters think they are looking to find, but it is the real world of the canal system that began to develop here two centuries ago. Take away the last reminders of that world and you take away the essence of the canal.

One can see why holidaying boaters might not rush to include the section of canal between Ellesmere Port and Chester in their itinerary. Lock-free, it heads through a flat landscape, dominated first by heavy industry, then by motorways and their complex of interchanges. It is not even always possible to use it at all – some years ago I tried to get through but found a Mini dumped in the canal and blocking up a bridge. Up ahead, however, lies Chester and the promise of more romantic scenes. They soon appear, though there is a little hard work first. The canal passes the old Tower Wharf and the arm leading down to the Dee, before turning smartly round a corner and clambering up through the Northgate staircase of locks. Beyond that it takes on the character of a moat to some vast medieval castle. Down in a deep, dark rocky cutting it runs beneath the lowering city walls – a part of the city, but self-contained, lost from sight in its own little world. Then the modern world reasserts itself. Where once wharves and warehouses dominated the scene,

Joe Skinner, whose boat
Friendship *is one of the centrepieces of the Ellesmere Port Boat Museum. He was unusual among boatmen in preferring mules to horses.*

now the supermarket rules – but Chester still has signs of industries past and present. A tall building proudly proclaims itself to be a 'Steam Mill', while the even taller shot-tower acts as a marker to the lead works.

Once out in the country, the canal 'of peaceful glidings and gently lapping water' of the tourist brochure can really be said to begin. It makes its way with little difficulty through a broad valley. When the canal was built there would have been little trade beyond coal for the countryside and towns, and produce heading back for Chester and for trading out through the Dee. Industry then meant little more than the corn mill. Christleton Mill shows the same characteristic patterns as the old warehouses back at Ellesmere Port. There is an industrial vernacular in architecture just as there is a domestic vernacular. There are contrasting materials – the rugged sandstone of the base, supporting the main structure in brick, its warm colours again contrasting with the hard coldness of iron. The straight lines of loading bays are balanced by the gentle, segmented arches of windows. The mill has now been converted for housing, but its character remains essentially unchanged. What one senses is a fitness for its purpose, just as when, at Beeston, one looks across to see the dramatic outline of the ancient castle on its craggy promontory. But the castle was no more designed to appear romantic than the mill was. It was built there because it commands a large area of countryside, and the hill gives it natural defences. It is essentially a functional building. It is only recently that we have learned to think of castles and mills as having any shared characteristics at all. It is perhaps as well that we have come to appreciate the visual delights of the latter – there are rather more of them along the canal than there are mouldering fortresses.

The canal contains many memories of its own history. Beginning life as the Chester Canal, it became part of a larger system once the Ellesmere Canal was built and, in time, that canal too became absorbed into a yet larger grouping. It became part of the Shropshire Union Canal Company, which in turn was to add railways to its interests, in a brave attempt to run a unified service. Ultimately it failed, falling entirely to the railway interest when the whole concern became part of the London & North Western Railway. Even that was not the end of the story, but it takes it through as far as the twentieth century. So the Chester Canal was to have many owners, but it was to be the Shropshire Union that went in for some major improvements. You can still see their work, notably by the locks at Bunbury. At the bottom of the locks, the neat little wharf building announces in fading letters that the 'Shropshire Union Railways and Canal Companies Carriers' will take goods all over the country. It is, alas, no longer true, but the building is still home to a canal carrier – a carrier of human traffic. It is a hire company office. And the building still shows its old role in the working life. It is constructed of brick but the corners have been rounded off and

Swing bridge, Crinan Canal.

reinforced with sandstone, so that any carts delivering goods would, with luck, bump round the corner rather than through it. The principal improvement brought about by the Shropshire Union Company was the provision of a fine range of stables beside the locks, which would not disgrace the most prestigious racing stables.

At Barbridge, the junction appears that the old Chester Canal Company had always seen as the most important part of their canal, the section that would let them in to the rest of the canal world. It arrived half a century late. The arm across to Middlewich on the Trent & Mersey Canal was not opened until the 1830s, when the whole line had been extended south by the Birmingham & Liverpool Junction Canal. It made handsome amends for its late arrival by presenting the canal scene with one of its most graceful bridges. The Middlewich branch is spanned at the junction by a long, low bridge that carries the towpath of the main line. It rises so slowly on such a gentle curve that it scarcely appears to disturb the horizon, only the sparkling white of its paintwork forcing it on one's attention, as it stands out in bold relief against the grass. Beyond that an altogether more modest junction signals the arrival of the Llangollen Canal.

From the first that special appeal that has won over so many, and turned casual holiday-boaters into life-long enthusiasts, makes its presence felt. At the same time something of the specific canal nature of the route becomes apparent – and the first comes in large measure from the latter. The chief engineer, as on the Grand Junction, was that key man of the mania years, William Jessop. He it was who, after many arguments between the different factions backing different routes, laid out the line of the canal and prepared the working drawings that would be the basis for construction. But when one looks at Jessop's commitments at this time, it is clear that this is all he would be able to do, other than keep an overall view of the works and be available for consultation if needed. What was required was, as the official title had it, a 'General Agent, Surveyor, Engineer, Architect and Overlooker of the Works'. This work went to a man, then 36 years old, who had slowly worked his way up from an impoverished childhood as son of a Scottish shepherd, who had died when he was still a baby. Apprenticed as a stonemason, he found work on some grand buildings, including Somerset House in London, and by 1786 was Surveyor for the County of Shropshire. This man was Thomas Telford. How much of the credit for the canal should go to Telford and how much to Jessop? At this distance in time it is not easy to say with precision: that the overall pattern was Jessop's is clear, but some of the architectural flourishes undoubtedly can be put down to Telford – his first love was architecture rather than engineering. The details are really of little importance, for this is a canal where everything fits seamlessly together.

It begins with a flourish, a flight of locks that lifts the canal up from the Cheshire plain and sends it on its way. With each watery step, a little more of the view opens out of a landscape that is, in its way, as much an invention of the eighteenth century as the canal itself. On every side the land is divided up into neat, squared-off fields, separated by equally neat hedgerows, with just the occasional tree to add variety. This is the landscape of enclosure, where the old common land was parcelled up and handed out to 'improving' farmers, for more efficient use. Crops improved, cattle and sheep grew fatter, and the countryside could meet the demands of the growing towns. Without the agricultural revolution, there would be no Industrial Revolution – and without these, there would have been no canal. The view from Hurlestone locks is like a wonderfully detailed illustration for an immense history book. The peaceful canal, the now quiet waterway, and the distant industries of Ellesmere Port are all tied together.

There is a wide view to delight the traveller, and a narrow one. One mistake Jessop seldom made was to underestimate the need to provide water for the canal. Alongside the locks is the reservoir that keeps the locks and the canal above them amply supplied. The high banks dominate the scene, but the locks themselves offer familiar pleasures – and a rare assortment of paddle gear. At the top is a modest house, four-square with no refinements beyond the little bay windows. The bridge is rather more colourful than the house, with alternating courses of red and blue brick giving a striped effect and, a neat little note, an arch built into the abutments to hold stop planks. Alongside it, the modern road bridge appears as a featureless, drab concrete slab.

The canal now makes its way on a comparatively trouble-free course, with a few twists to cope with gentle undulations and a spatter of locks as it begins its gradual climb. For many this is ideal holiday-boating – rural surroundings, fine views and just enough locks to keep the children busy, but not so many as to turn them mutinous. Everything along the way speaks of a canal serving an essentially agricultural community. There are no attempts to get to the heart of communities. Wrenbury wharf, for example, stands well outside the village but has developed its own little cluster of buildings – mill, houses and pub with, at their centre, one of the lift bridges that are very much a feature of this canal. The bridge has a superstructure with an overhanging beam, which counterbalances the weight of the bridge platform. In the smaller bridges, the platform is raised by hauling on a chain dangling from the beam. This is fine until the chain becomes detached: as anyone who has ever tried it can testify, lifting one of these bridges without a chain is fiendishly hard work. These are the types of bridge familiar from the paintings of the Dutch canals by Van Gogh, but they were not built to create picturesque landscape features. The best engineers, even in the early 1790s when money appeared plentiful, were cost-conscious – and none more so than Jessop.

A lifting bridge is a confounded nuisance on a busy road, but quite acceptable if all it is required to do is provide an occasional route to enable a farmer to take his cattle from one part of his land to another. A glance at the older houses in the area shows that they are timber-framed, with the gaps originally filled by wattle and daub. This suggests at once that local builders found brick and stone to be expensive luxuries – and the same would have held for canal-builders. The moveable wooden bridge was cheap to build. True, it was more expensive to maintain, but those were costs that could be met out of what it was hoped would be a very respectable revenue. Lift bridge and cottage were both responses to the simple question – what are the cheapest local resources?

As the boater moves on towards Wales, so the horizon up ahead is increasingly dominated by hills and the pace of travel changes. Locks appear with increasing frequency until, at Grindley Brook, there is a sudden charge up the hill – three locks close together, followed by a three-lock staircase. Here the ridges that provide footholds beside the lock set up a rhythm like that of the op-art paintings popular some years ago. The ridges climb the slope in diminishing perspective, to be repeated a little closer together, it seems, with each successive rise. And it was here that Telford allowed himself one of his little architectural flourishes. The lock cottage has a bow-front, around which is spread a charming little verandah.

Having hauled itself up the hill, the canal now has to squirm round a hill to maintain a level. While such extreme contour-cutting would have been unremarkable on a Brindley canal, it sits uneasily with the notion of a canal of the 1790s. There was method here, however, for the bend took it closer to Whitchurch and shortened the branch that was built to the town centre. Time and again one is struck by the careful planning that went into this route. It is not always obvious from a boat, but travel alongside the canal by road or walk the fields nearby and you will see that what appears to be simply a canal making its way through the countryside is actually being carried on a series of low banks, and is cutting almost imperceptibly into the hill. It is a matter of fine balance to choose between a detour along a contour and expensive, even if slight, earthworks. Here it is all done with great subtlety, which helps to create the sensation of moving effortlessly along an apparently wholly natural route through the countryside.

This sense of canal and landscape working together ends at Whixall Moss. When the last glacier retreated from this area at the end of the Ice Age, it left hollows filled with water. Some remain as lakes or meres, but others became clogged with vegetation that rotted down to form peat bogs. Whixall Moss was one such bog, a black, oozing wasteland. Here the canal was cut straight through the middle, left at first as a drain, then built up as a navigable canal carried on a bank. The reclaimed land remains on one side with its rich, deep soil; to the other there is still a scrubby waste. This is one of those rare cases where the canal produced a fundamental change in the landscape

Soulbury, Grand Union.

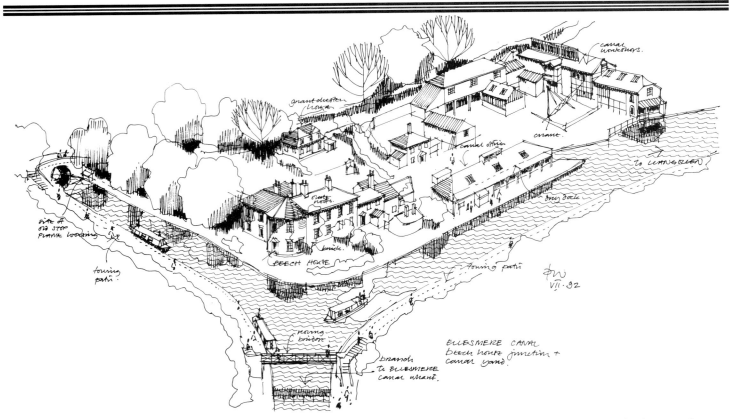

Now most commonly known as the Llangollen Canal, the Ellesmere Canal began at Hurlestone Junction, where it connected with the Chester Navigation, and passed via Ellesmere to Frankton and then down to Correghofa, where it joined the eastern branch of the Montgomeryshire Canal. The canal to Llangollen was originally a branch line. At Ellesmere Junction, where a short branch leads off to the town wharf, the fascinating company canal yard complex, with the splendid Beech House, creates an imposing junction 'feature'. Dry docks, workshops, forges and lock-gate sheds make up an amazing collection of yard buildings, and this important part of our industrial heritage and canal archaeology needs interpreting and managing with imagination.

itself. This brief straight-line interlude contrasts with what has gone before and even more with what lies ahead, for the area round Ellesmere is one of hump and hollow, through which the canal must struggle to find a way. Many of the hollows are water-filled, creating a miniature Lake District. The meres are a delight and the canal goes close by, so that at times only a few trees separate the boat from the wider expanse of water. And beyond that lies Ellesmere itself, the town that gave the canal its name.

Ellesmere is still very much the heart of the canal system. Here is the maintenance yard – a curious place, with its half-timbering, giving it not so much the air of a workplace as of a piece of misplaced stockbroker Tudor, floated over from Surrey. Here too is the company headquarters, Beech House, where Telford repeated his bow-front theme from Grindley Brook, but on an altogether grander scale. A short arm leads down to the town wharf with crane, warehouse and creamery – Ellesmere was a centre for cheese-making long before the canal came along. Ellesmere is, indeed, a sort of summation of everything that has come before: the market town, the centre for a wholly rural area, where the nearest thing to an early industrial building is the 200-year-old rennet works. It reaffirms the sense of a canal that allows one to drift lazily through unspoiled country. That sense will remain for the rest of the journey, but in reality the canal of two centuries ago was thrusting into a highly industrialized region.

Beyond the Ellesmere wriggles lies Frankton Junction. Here the Llanymynech branch heads south to link into the Montgomery Canal. Abandoned years ago, there are now new signs of recovery. A recently cast mile-post announces that there are 35 miles of canal lying ahead before the end is reached at Newtown. Restored locks show what has already been achieved, and a neat, unpretentious lock lobby stands at the top of the flight. The main line continues its gentle, somewhat meandering way with little to hint that anything much has changed in the countryside that lies all around. But when the canal was new this was a busy coal-mining region. Nothing, however, can disguise the fact that the hills are getting ominously close and that further progress is going to require something more than a wriggle of the hips to squeeze round the obstacles. It is not so much the hills that create the problem but the two valleys that come through them – the Ceiriog and the Dee. These were daunting obstacles for any engineer and before any decision about the crossings was taken, the resident engineer, Thomas Telford, was called away for a brief period to work on another canal.

The Shrewsbury Canal was another of the group that received its enabling Act in 1793, but it was a comparatively modest affair, linking Shrewsbury with the Ketley Canal, with its inclined plane. It was built to serve the local iron works, and the local iron master William Reynolds took a keen interest in its construction.

A tub-boat loaded on to its cradle at the foot of the Trench incline on the Shrewsbury Canal. The winding engine house can be seen at the top of the plane.

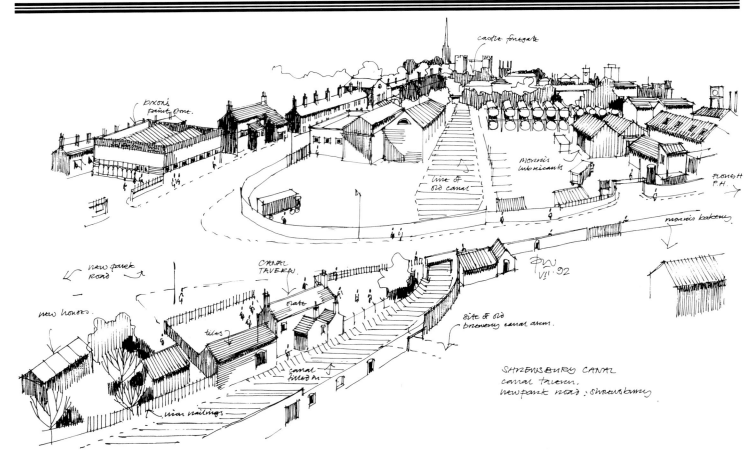

Labels in image: castle foregate, dixon's paint store, morris's lubricants, line of old canal, plough P.H., morris's bakery, new park road, CANAL TAVERN, slate, new houses, tiles?, canal filled in, site of old brewery canal arm, iron railings, SHREWSBURY CANAL canal tavern. newport road, shrewsbury

Linking to the Newport branch of the old Shropshire Canal at Wappenshall Junction, the Shrewsbury Canal used to run via Wombridge and Berwick to the Castle Foregate basin in Shrewsbury. At Wappenshall there were links to the Shropshire tub-boat canal system and the Trench inclined plane. The main Shrewsbury Canal was 14½ miles long and the link to the tub-boat section three and a half miles. Surmounted by a demolished brewery, new houses and a paint store in Shrewsbury, the Canal Tavern stands next to its old 'filled-in' canal, which once went on to Castle Foregate basin. The smell of lubricants and of a bakery fills the air.

Most of the canal has now disappeared. In Shrewsbury itself, the Canal Tavern still stands, but all that remains around it are walls that were once parts of warehouses and a shallow, rubbish-filled hollow. The engineer was William Clowes, and for the first 12 miles the work offered few difficulties, running along on the level. The last four miles included locks, a tunnel – one of the first long tunnels to enjoy the luxury of a towpath – and an aqueduct across the River Tern at Longdon. Work had just begun on the latter when Clowes died and the company borrowed Telford to see the work through.

Telford and Reynolds seem to have been men who relished the idea of innovation, and Telford himself was struck by the structural possibilities that presented themselves with the use of iron. It has to be remembered that it was in this region that the world's first iron bridge had been built across the Severn, and where the first iron tub-boats had been tried out on a canal. But the metal was finding new uses all the time. In that year, two engineers hit on the notion of applying it to aqueducts. Benjamin Outram seems to have been first away, if only by a few weeks, when he designed an aqueduct for yet another 1793 canal, the Derby, but the important decision was that of Reynolds and Telford to use iron at Longdon-upon-Tern.

The approach to the River Tern is via a series of brick arches, suggesting that Clowes had no notion of producing anything other than a conventional aqueduct, no different from those of the Brindley age. Then iron takes over in a somewhat crude, spindly structure. The trough is carried on triangulated iron supports, with the towpath alongside. No one could describe it as beautiful, or indeed even slightly attractive, but that is not the point. It worked. The old-style brick and masonry aqueducts had to be built to massive dimensions to withstand the pressure of water and the weight of puddled clay needed to keep the water in. With the iron aqueduct, the metal plates alone could take all the pressure. If an engineer needed a long, high aqueduct, then iron was the perfect answer – and that was exactly what was required for the crossing of the Dee valley.

Telford went back to the Ellesmere Company fired up with enthusiasm for the new technology, and his message fell on receptive ears. Jessop was already in partnership with Outram at the Butterley Iron Works in Derbyshire and knew of Outram's own success on the Derby Canal. He at once gave his wholehearted support for the idea of a cast-iron trough to be carried on tall piers across the Dee. The notion of a long aqueduct was already his first choice, and he agreed with Telford that iron would be the obvious material to use. The plans went forward and the result was Pont Cysyllte, the greatest monument of the canal age.

Longdon-on-Tern aqueduct, the very modest forerunner of the greatest of all iron-trough aqueducts, Pont Cysyllte.

Everything is on a grand scale. There are 19 arches, their piers built of solid masonry up to a height of 70 feet, then left hollow and braced by cross-walls. Above them is the iron trough 1,007 feet long and, at its highest, 121 feet above the Dee. The towpath is cantilevered out over the water, leaving a narrow gap just wide enough for one boat. Travelling across was probably the nearest most boatmen ever got to flying, for the edge of the trough was below the gunwales of the boat, so that one could quite easily step out into space. If the keynote of the Llangollen Canal is calm, peace and quiet, then this is the point of contrast, offering a sense of excitement and a frisson of danger – even if the danger is more in the mind than present in reality. Pont Cysyllte is so dramatic that the neighbouring Chirk aqueduct, which would be the four-star attraction on almost any other canal, seems almost tame. Yet it too partook of the new technology – only here the iron trough is less obvious, simply acting as a lightweight lining for the solid masonry. Chirk also suffers from a further disadvantage: it is overshadowed by the viaduct built by the Shrewsbury & Chester Railway. Even the Board of Trade inspector, Captain Wynne, was saddened, remarking that the viaduct 'completely degrades' the aqueduct, and adding that 'The two are so mixed up by their close juxtaposition that the proper effect of each is lost, and the scenery which is very beautiful is not improved.' It is hard to quarrel with his verdict.

In the excitement of the two river crossings, it is easy to lose sight of the true significance of this part of the journey. At Chirk aqueduct one can look down on a large water mill far below beside the Ceiriog; what is less easily spotted is the basin at the end of the tunnel, where a tramway once came in, carrying slate from the mines in Glyn Ceiriog. At first it was worked by horses, but in later days steam locomotives took over. At least when crossing Pont Cysyllte, if one can keep one's eyes off the drop down to the valley floor, the signs of industrial life are still there in the shape of the factory chimneys. Once across the aqueduct, the canal bends round to head up the Dee valley towards Llangollen, but a short arm leads straight on to a somewhat inconsequential end. Yet this arm, now housing a hire-cruiser base, was the key to the success of the whole canal. From here, a railway ran on to Ruabon, to an area rich in coal mines and iron works. It was here, at Plas Kynaston, that the parts for the Pont Cysyllte were cast – here, too, that Telford came to order castings for other iron bridges as far away as the Spey valley in Scotland. Now the only sign of this industrial past that is visible from the canal are the odd chunks of furnace slag that can be picked up near the towpath.

The tourist route makes a splendid conclusion to the journey. This is the narrow, navigable feeder, hugging the side of the steep hill slope, and faithfully following every natural bend in the land itself. It is justly famous for its beauty, weaving a way through the trees, then opening out to give majestic views of limestone crags and the romantic ruins of Castell Dinas Bran, the ancient fortress on its high hill. Yet even here the memories of an industrial past linger. A bridge has a double arch – why? The answer is that the second arch carried a tramway, which brought stone down from quarries in the hills. Much of this history remains unnoticed by the thousands who come this way each year, but it is surely the most remarkable thing of all about this remarkable canal. Here is a route designed to meet the most mundane needs of a workaday world – humping coal and iron around the country in slow-moving cargo boats – yet two centuries later it is admired as an escape from the same world it was built to serve.

What would the planners, the engineers, the boatmen, the navvies think of the canal scene today? Could they ever have envisaged that the route, which to them represented transport technology in its modern guise, would be treasured purely in terms of leisure and pleasure? Even as early as the promotion of the Trent & Mersey Canal in the 1760s, pamphlets were issued praising the delights of having a 'lawn terminated by water', but this was more of a hopeful suggestion to landowners along the route than a serious plea to householders. But people were travelling for pleasure from the earliest days: trip-boats went from Chester to Ellesmere Port to allow passengers the somewhat dubious pleasure of having a swim in the Mersey. This, however, represented the canal as a means of transport: there was no indication that the journey was enjoyed for its own sake.

I first travelled this route some 30 years ago, when canal travel for pleasure was still a novelty. The first hire fleet had only recently been established and the first cruising guide had just been issued. I knew nothing about canals and – it soon transpired – even less about boat handling, but an enthusiasm was born that Summer that has never died. It is tempting to say that I recognized from the first that it was the special character of the working canal that registered, that made this waterway so different from the rivers I had tackled by canoe, but it is not really the case. But the fact that I made no attempt to analyse the popular appeal of the waterway is not to say that it was not that special character that brought me back time and again over the years. Few of us bother to analyse the pleasures of a holiday, but unless someone does so, then those elements can all too easily be unthinkingly discarded. The character of the Ellesmere Canal was created in the period in which it was built to serve an industrial society. It grew out of conscious decisions by men such as Jessop, whose careful, economical planning fitted it so perfectly to its natural surroundings; and Telford, who loved the occasional touch of bravura. The canal has changed and developed over the years, but the essence remains – the character that sets it apart from another canal of another age. The Ellesmere Canal has its

ABOVE: *Mooring ring, Crinan.*
OPPOSITE: *Lock entrance, Sharpness.*

own individuality and personality. Compare it, for example, with a later canal – and one for which Telford was the engineer – which was to be joined with the Ellesmere Canal as part of the Shropshire Union system.

The Birmingham & Liverpool Junction was begun at the very end of the canal-building period, in 1826, a year after George Stephenson had driven his engine Locomotion 1 at the triumphal opening of the Stockton & Darlington Railway. The techniques used in its construction were very much those to be used in the great rush of railway construction that was shortly to engulf Britain. 'Cut and fill' is a very good brief description of the methods. Where the canal met a hill, it was driven through in a deep cutting; where it came to the valley beyond the hill, the spoil could be used to build an embankment. To the canal traveller it is the great cuttings that remain in the mind – in summer, dark green shady tunnels of overhanging trees; in winter, an eerie place of cold shadows and drooping, twisting tendrils. There is no mistaking the artificial nature of these cuttings, with their tall bridges arching high overhead. The banks are less immediately impressive, but were to prove infinitely more troublesome to the engineers, constantly shifting and resisting all attempts at stabilization. The most difficult was Shelmore Great Bank, and looking at a modern Ordnance Survey map it seems wholly unnecessary – the canal curves away to the west as if deliberately seeking out the low level of the valley, when by keeping to a straight line it could have kept almost to a level. The answer is to be found to the east of the line: Norbury Hall and its surrounding park. The local landowner had enough influence to keep the unwanted canal at arm's length – and how Telford must have cursed him. When it was finally opened to boats, the rest of the canal had long since been completed – and Telford himself had been dead for six months.

The Birmingham & Liverpool shows its character in detail as much as in overall plan. The way in which the canal was pushed relentlessly through the countryside, ignoring, it seems, all natural obstacles, is vividly demonstrated in the lovely little Tyrley flight of locks. Descending them, one goes from open country to a little man-made ravine, where the sandstone walls still bear the marks made by the drills as the workmen prepared to blast the rock away. By now the use of iron aqueducts was no longer a novelty – and another advantage had appeared. Once you had a pattern made, it could be duplicated. So the same castings were used at Stretton and Nantwich, and were to turn up again in another late canal, the Macclesfield. The Stretton aqueduct is the grandest, with finely carved stone pillars at either end. Perhaps it was Telford's way of tipping his hat to an earlier generation of engineers – for the road it crosses started off as Roman Watling Street. Equally it could have been self-congratulatory, for earlier Telford had been appointed to survey the road

The stone cutting at the foot of Tyrley locks on the Shropshire Union still shows the marks of holes drilled for blasting.

from London to Holyhead, and although he had no need to improve on the Romans' arrow-straight alignment, he had a proprietorial pride in the road as a whole. Other Telford touches appear along the way. The lock cottages are consciously designed, with their low pitched roofs and frontages enlivened by the use of arched recesses in the walls. These small touches bring constant interest and delight to travel on the canal, but what one remembers, what comes instantly to mind when the name of the canal is mentioned, is the drama of bank and cut, the sudden change from floating high above the surrounding countryside to being sunk down deep in the enclosing walls of the cuttings. The personality of this canal is as different as can be from that of the Ellesmere, which would seem like a country lane to the other's main road, were it not for the even greater excitement of Pont Cysyllte.

One reason that people return to the canals year after year is that no two canals are ever quite the same. It is not merely a matter of going through different sorts of countryside, it is just as much to do with the canal's individuality. The countryside of Shropshire is not so very different along the Ellesmere Canal as it is along the Birmingham & Liverpool, but the experience of going through it could scarcely offer a greater contrast. This is admittedly an extreme example, but when one turns to another of 1793, the Brecon & Abergavenny, one finds a canal as wholly distinctive as any in Britain.

Most authors who write regularly about canals will find themselves being asked the same question: 'I want to go on a canal holiday – where should I go?' There is no obvious answer, because it depends what the questioner is looking for in the holiday, but when one tries to find out more, the same set of requirements often appears. People want 'to get away from it all' – 'all' in this case being the rush of traffic, the noise and bustle of towns. They want attractive scenery, peace, quiet and – almost inevitably – 'not too many locks'. To the beginner, the lock can be somewhat intimidating, not to say incomprehensible.

Few canals come closer to this ideal than the Brecon. It not only passes through beautiful scenery, but runs throughout its length within the Brecon Beacons National Park. It seems tailor-made for those who want little more than a quiet chunter, with nothing to disturb them and nothing whatsoever to remind them of the world of work that for a week at least they simply want to forget. But the Brecon was no more built for leisure than was the Llangollen, and those with a taste for history will find that this perfect rural canal serviced one of the heartlands of the Industrial Revolution. The wording of the Act for what was then called the Brecknock & Abergavenny Canal puts it into perspective:

> An Act for making and maintaining a navigable Canal from the town of Brecknock to the Monmouthshire Canal near the town of Pontypool, in the county of Monmouth; and for making and maintaining Railways and Stone Roads to several Iron Works and Mines.

If one were to give that description to the holiday-makers, they would probably find it less than enticing – iron works and mines? No, thankyou. Yet in good measure the fascination of this canal lies in the fact that one can cruise it quite happily with no thought at all about industrial history – yet all the clues are there to be followed up by those who wish to do so.

From the first, the canal was built to connect with the Monmouthshire Canal, which ran down to the River Usk at Newport. The two waterways met at Pontymoile just outside Pontypool, where a toll-house was built to make sure that each company received its due fees. It was a very friendly arrangement as far as the engineer was concerned – Thomas Dadford Junior of the Brecon & Abergavenny dealt with his opposite number on the Montgomery, his brother John. Pontypool itself certainly qualifies as a major industrial centre, but very little of this aspect of the town is visible from the canal. Indeed, the most visible reminder of the industrial past is one stage removed from the dirt and grime of foundry and pit. Pontypool Park surrounds a fine house begun in the seventeenth century, which became home to the Hanbury family, founder of the local iron works. Today, the stable block has been converted to tell the story of the area – but without that exhibition there is little to distinguish this from any other park and any other country house.

Split bridge, Kidderminster, Staffs & Worcester.

The Brecon & Abergavenny Canal at Pontymoile.

Once clear of Pontypool, the canal sets off on what will be an increasingly tortuous journey. It was to be tortuous in other ways, too. Although by now there was a well-established system of contracting out canal work and a large body of experienced contractors, work on the Brecon went to an odd collection of locals, covering every social status from William Watkins, labourer, to William Parry, gentleman. Progress was slow, and a familiar story was repeated – money ran out, more was raised and the work staggered on until the grand opening finally arrived in 1812, after a rate of progress (if progress is the word) that worked out at less than two miles a year.

There was little doubt about the direction the canal would take: the geography of South Wales has always dictated where its major transport routes would run, whether road, canal or rail. The valleys all run roughly north to south, separated by tall ridges of hills. The canal set off to follow the edge of the ridge that ends in the shapely green hill of Blorenge and stuck tenaciously to its contour: if a valley cut into the hill, then round the little valley the canal must go, even if it involved an extravagant horseshoe bend; and if the hill bulged out towards the river valley, then once again the canal must follow suit. It is in many ways similar to the final section of canal between Pont Cysyllte and Llangollen, offering the same high-level route and, if anything, even more spectacular views. This is particularly true as the canal swings round the flank of Blorenge to turn west to follow the Usk valley towards Brecon. In spite of its name, it never actually reaches Abergavenny, but it does offer superb views of the town under the shadow of the tall cone of the Sugar Loaf. It then continues on its way, lock-free for 23 miles – one of the factors that helps make this canal so popular. Then a flight of five locks drops it 50 feet down to the Usk valley, which has been gradually rising to meet it.

Theoretically, life should now have been easier for the engineers, and so in a way it was, but in its upper section the Usk valley becomes ever narrower, so that the canal has to disappear briefly into a tunnel as a spur of the hill pokes right out to the river's edge. Eventually the encroaching hill pushes the canal off its perch, and the Usk is crossed on a masonry aqueduct at Brynich. After that comes just one more lock and a trouble-free run to Brecon – or it would be trouble-free, but for changes at the terminus. Once boats stopped at Brecon basin, turned round and came back again. Now the basin has been filled in so that boats venturing to the very end find that there is no longer anywhere at all to turn round and they have to come out backwards for quarter of a mile until they reach a winding hole and can resort to a more normal form of progress.

The sense of travelling a wholly rural canal is scarcely interrupted throughout the journey. True, there are wharves along the way, such as that at Govilon, which speak of trading days, but very little to show what that trade might have been. This is again a direct result of that special geography of the valleys. It would have been hopelessly uneconomic to try and take a canal up into the hills, and even more absurd to try and put a canal on an east–west alignment over the high ridges. This was where the 'Railways and Stone Roads' came into their own. The Dadford brothers set out to construct the rail route that would go where no canal could pass.

Thomas Dadford Senior had worked on the Trent & Mersey Canal when the Caldon branch was being constructed. As a canal, it rivals the Brecon for its peace and beauty. It too follows a river valley, that of the Churnet, and here at least there are real signs of industrial life. At Cheddleton there are a pair of water mills, designed not for grinding corn but for grinding flints for the local pottery industry, where they were mixed with the clay to produce a lighter-coloured body for the pots. The works are very sophisticated, designed to save labour and to have process follow process as smoothly as possible. The flints themselves were brought by boat round the coast from East Anglia and along the south coast, and then up the canal to the wharf. The first stage in the process involves calcining, heating in a kiln to make the flints more brittle and easier to work. Because the canal is higher than the river, the kilns were built into the wharf itself, so that the flints could be unloaded straight from the boats to the furnace, and after heating could be taken out at the bottom and away on rails to the mills. Here, mixed with water, they were crushed under rotating arms loaded with heavy stones. The slurry was dried and

OVERLEAF: *The Great Glen, Caledonian Canal.*

The River Churnet and the Caldon Canal meet at Consall Forge, a beautiful rural spot that was once a busy industrial site.

the powdered flint sent back by canal to the potters of Stoke. In later years, steam power took over from water power, and the steam-powered Etruscan Bone and Flint Mill stands at the junction of the Caldon and the Trent & Mersey canals. The mills, however, were rather in the nature of by-products of the canal, taking advantage of the new transport system. The canal was built in the first place to reach the stone quarries of Caldon Low. The quarries were higher up in the hills, so the route was continued by a tramway: the rails consisted of flat iron plates laid on top of timbers – in effect, a wooden railway with an iron top surface for extra durability. Traces of the old route can still be seen leading away from Froghall basin at the end of the line. It makes for a pleasant walk, with a little detective work thrown in for those who want a break from boating: and if they find that an entertaining pastime, then they will find a visit to the Brecon & Abergavenny a positive feast of pleasure.

Where the Caldon was continued by a single railway, the Brecon offered a very different solution. Here a whole series of tramways, over a dozen of them, fed down into the canal. Their construction was a little different from that of the Caldon, closer to that of the conventional railway we know today, with iron rails carried on sleepers, but because the work was done by horses there had to be a clear path down the centre. As anyone who has ever walked down a railway will know, the sleepers make this very difficult, so on the tramways the rails were mounted on two parallel rows of stone blocks. This gives the tramway detective his clues, for the stone blocks were often left in place, where wooden sleepers would long since have rotted and decayed.

There are no records to show just how many people became canal enthusiasts through a first introduction to the Llangollen Canal, but the number could well run into many thousands. The canal seems to offer so much to the pleasure-boater. It is a wholly rural canal, passing through delightful countryside, but it has its moments of – literally – high drama, with a grand aqueduct at Chirk, followed by the even grander crossing of the River Dee on the Pont Cysyllte. Many holiday-boaters are happy to leave it at that,

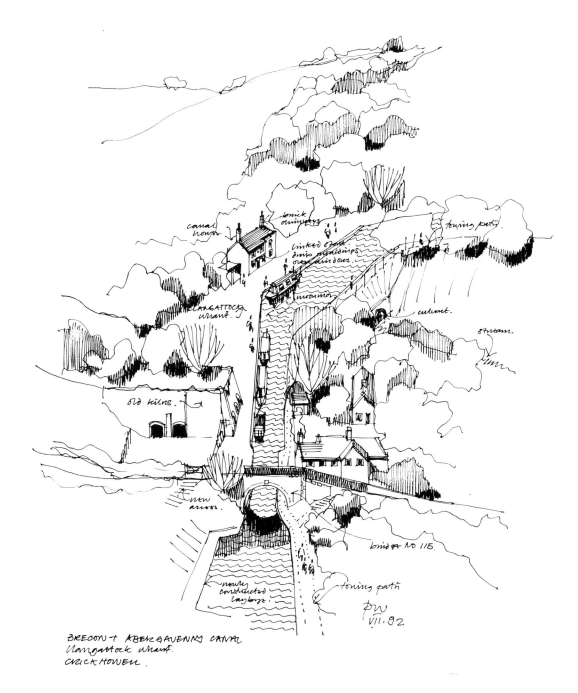

BRECON + ABERGAVENNY CANAL
Llangattock Wharf.
CRICKHOWELL.

Now popularly known as the 'Mon & Brec', this waterway began at Brecon basin and extended for 33 miles, via Llangattock, Govilon, Llanfoist and Mamhilad to Pontymoile, where it formed a junction with the Monmouthshire Canal. Stone cottages with slate roofs, stone bridges and lime kilns are reminders of the materials once carried on the canal. Fuel and stone were loaded from the tramway that ran along the top of the kilns, and lime was taken away at the bottom by boat. The canal itself is, in many places, cut into the side of the hill, the excavated material having been used to form the towpath. The compelling combination of industrial archaeology and fine natural landscape combine to make the Mon & Brec one of Britain's most beautiful canals.

still be found, which not only show where the line went, but also indicate that this line was built to a 3-foot 6-inch gauge. Many of the tramways have similar relics, but none is so packed with interest as the one known as Hill's tramroad, named after the Hill family, iron masters of Blaenavon. The first part of the story begins at Llanfoist.

The first indication that something out of the ordinary is happening at this spot comes with the canal bridge: where most are hump-backed, this one is flat. Those who stop to investigate further will find that this is not the only canal crossing at this point – there is also a somewhat dank, dripping tunnel under the waterway. Then there is a wharf area, with quite a grand house and an unusual warehouse. The latter is a two-storey building, where boats can float in underneath and the roadway appears at the far side of the building at first-floor level. The final piece of the puzzle appears in the form of a track, running straight up the hillside behind the wharf. Put all the elements together, and it makes sense.

The explanation begins with the hillside track. Walk just a little way up the track and you will find the stone sleeper blocks of a tramway; and those with energy to spare can climb all the way up to the point where a platform was cut for a brake drum. So this was a tramway incline. Returning to the wharf area, it is then obvious that at the foot of the incline the track must have divided. One branch ran down to the warehouse, where goods were unloaded into the upper level ready to be lowered down to the boats waiting underneath – a small-scale version of the old Telford warehouses that once stood at Ellesmere Port. The other branch went on across the bridge and down the hill to the village of Llanfoist – hence the flat-topped bridge, and hence the tunnel, since no right-minded pedestrian would want to share a bridge with trucks that had just thundered down a steep slope. All this makes sense, but it still leaves one basic question unanswered. Why was the system built at all – what industry did it serve? The answer to that requires a little more investigation. An energetic walk up the track of the old incline brings one eventually to a broad ledge running round Blorenge hill, and following that leads one to a complex meeting of the ways. One section runs off across the road from Govilon towards the quarries at Tyla, but another section heads south to Blaenavon. It is here that the whole system really begins to make sense. This is industrial South Wales, and not just that, but industrial South Wales of the canal age. An eighteenth-century iron works, with massive stone blast-furnaces rising like the towers of a medieval castle, stands not far from the Big Pit colliery. The working life of both has ended – and both are now preserved as museums – but one can still see just how grand and important they once were. This is the world served by the Brecon & Abergavenny Canal. It was for this that the engineers and financiers struggled to get the canal completed. There are perhaps few canals where the direct link between heavy industry and canal transport is so strong and, paradoxically, there are few canals where the holiday-maker can more completely feel that the everyday world of work and industry has been left far behind.

The reason that the Brecon Canal is so immediately appealing is not difficult to find: the tramway connections kept industry and waterway apart. Now that the tramways are no more, only those who actually want to make the connection will ever be aware of it. For many, however, the historical connections, far from detracting from the enjoyment of travel actually add to it. Even on those canals that show their true colours, openly and boldly, the appeal can be immense. Nowhere is there a better example than on one of the canals that only just made it into the canal age at all – the Macclesfield. There was a great deal of discussion among the proprietors when they were preparing their plans for the Parliamentary session of 1826. Which should they go for – the old, tried and trusted canal, or the new-fangled, and yet to be proved, railway? They decided to go for the canal, and although it may not have been the best option for their shareholders, for one canal enthusiast at least it was a splendid decision. Here, if anywhere, the delights of fine scenery and the fascination of historical connections combine to perfection.

The Macclesfield proclaims its modernity and independence from the start, leaving the Trent & Mersey at a gentle angle, then cutting across it on an aqueduct, much as a sliproad leaves a motorway. The overall plan is very similar to that of its contemporary, the Birmingham & Liverpool: locks huddled together in one long flight, the countryside attacked head-on through cutting and bank. Just as on the Brecon, it is the hills that dominate the line, the dark gritstone outlining the edges like the margins of Victorian mourning cards. The countryside through which the Macclesfield passes is seldom softened, however, as the Brecon is by its dense woodland: stone rules here and dominates the canal itself just as it dominates the landscape. There is a special kind of satisfaction in seeing the way in which the rough, ragged stone outcrops have been tamed by skilled masons; how the blocks have been trimmed and shaped and fitted to form everything from road bridges to the setts that form overspill weirs at locks. The most famous examples of the mason's art are to be found in the so-called 'snake bridges'. These serve a wholly practical function. Wherever a towpath turns over from one side of the canal to the other, a special type of bridge has to be supplied. The horse cannot simply drag the boat across the bridge, so the simplest answer is to have the path loop back on itself to pass under the arch. Here the towpath climbs up, then turns in on itself in the most perfect curve, so that each turn-over bridge on the Macclesfield appears as a genuine work of art. One might compare worn bollards with modern sculpture, wonderful accidents created over time – but this is no accident, rather a conscious decision to create a bridge that does not just satisfy a practical need, but surpasses it. This is designing at its best.

If the hills provide the main theme of travel on the Macclesfield, then the mills provide its accompaniment. The small, northern mill towns do not give themselves away to the casual passer-by. They are closed in, rather secretive places. Bollington is a splendid example. The canal comes in at a high level, heading off towards the mills: grand, exclamatory buildings, proclaiming their importance in their size and architectural embellishment. But down below there is nothing exclamatory about the town. Houses huddle together; streets are narrow; and openings lead not to wider views but to ever more secret courtyards and alleys. Bollington is a good spot for nosey-parkers, who like to speculate what might lie just around the corner. It is a place that seems to be in awe of the surrounding hills and has turned in on itself for comfort. But at the same time there is the unity that comes from the stone – the stone of church, bridge, mill, aqueduct and cobbled yard. Anyone who fails to respond to the Macclesfield will probably never enjoy a northern canal.

Reaching the end of the Macclesfield is not quite the end of this particular story, for at its northern extremity it runs into another waterway, the Peak Forest, and offers a choice of ways – on towards Manchester, or away into the countryside. The Peak Forest was one of those that just missed the 1793 rush, receiving its Act the following year, and it too is a canal with a very strong northern, character. Turn north and one heads for a route at once full of drama and yet at the same time delighting with little details. Almost at once it plunges downhill on the long Marple flight of locks – and what a pleasure they are. There is a reassuring solidity about the great blocks of stone that appear in the boundary wall, in the lock chambers and in the little bridges that hop across the tails of the locks. Where the canal passes under the main road, the horse was supplied with its very own tunnel – horseshoe-shaped, not as a symbol of its horseyness, but because that is the most economic shape for an animal with thin legs supporting a fat body. The key building is the warehouse with its covered loading bay, for it was built for the local cotton magnate, Samuel Oldknow, whose prosperous business was to give the canal a good part of its trade. Wide pounds help to keep the locks supplied with water. Not so many years ago these locks were derelict and the reedy, muddy pounds attracted their fair share of mattresses, tyres and rusting bicycle frames. Now cleared out, shared with equanimity by boats and ducks, they have become prized features. House-owners who once tried to hide away behind high fences now look out with pleasure at these very superior side ponds.

The grand climax comes at the bottom of the flight when the canal has to cross the River Goyt. Marple aqueduct is immensely impressive, crossing 90 feet above the river on three tall arches. One trouble with big aqueducts, as with big bridges, is the sheer weight of the span itself. A solution was discovered by William Edwards of Pontypridd, who had immense trouble with a bridge he designed for the town. Three attempts failed; the fourth, with a huge span for a masonry bridge of 140 feet, succeeded. His solution was to reduce the weight in the haunches of the arch, where it sprang from the support, by piercing it with cylindrical holes. This device, known as a pierced spandril, was used here by the engineer Benjamin Outram. The holes are very decorative, but their job is purely practical. Marple aqueduct stands today as one of the finest monuments of the canal age.

Beyond Marple the canal comes ever closer to the old industrial heartland of Lancashire. The mills – if now rarely at work – appear with increasing frequency. At Dukinfield Junction, the Peak Forest gives way to the Ashton, a purely urban canal that carries the route on down to the very heart of Manchester. If the Brecon & Abergavenny is for many the ideal of rural bliss, then here is the exact opposite. The canals of Manchester bring you the world of industry and commerce. Mills,

A crowded construction scene at Piccadilly lock on the Rochdale Canal. The massive warehouses in the background are no longer there.

ABOVE: *Lapworth, Stratford Canal.*
OPPOSITE: *Journey's end, Dunardy, Crinan Canal.*

factories and warehouses rear up to create a man-made canyon in which the canal seems almost lost. At the very heart of the system is the Rochdale, so overwhelmed by its urban setting that one lock is now condemned to Stygian darkness, lurking beneath the foundations of a tower block. But if this is a world that some find unprepossessing and unattractive, it remains the world that the canals helped to create: this is what they were all about, not idyllic holiday-making, but the creation of wealth. Canals were the servants of industries that spread from here throughout the world. You glimpse something of this if you follow the canal system through Manchester itself. Names such as Bengal Street and China Street appear, grand buildings borrow their decorative devices from anywhere and everywhere. Here there is a touch of the Moghul styles of India; there a reminder of the earlier great trading empire of Byzantium. The buildings were sending out a message: they traded with the world as equals, and probably as superiors, though they would not wish to overstress the point. For some people, myself included, the world of the industrial canals is every bit as interesting as that of the rural, and you get a visual lesson in world architecture thrown in.

Those who still hanker after the quiet life need do no more than ignore the lure of Marple's locks and aqueduct and turn away in the opposite direction at the junction, to head for the end of the canal at Bugsworth – a name now gentrified into Buxworth. The canal now really does have something in common with the Brecon: the same high-level route along the side of the hill, the same wide views out over the valley. But where the Brecon keeps its industrial connections at arm's length, here they are very much on parade. The dark stone mills are an ever-present reminder of the canal trade, and at the terminus there is no mistaking the sense of of a once-busy world, where boats were loaded and unloaded by sweating crews. In recent times the complex of basin and wharves has been cleared out; the ruined lime kilns stand as proud as the blast-furnaces of Blaenavon; and in the hills the ragged skyline tells the story of quarrying. The Peak Forest offers scenic delight in plenty, but it never quite loses that direct visual connection with the world that brought it into being, and never lets us forget that our holiday route was once one of the main trading arteries of the Industrial Revolution.

The journey down the Peak Forest also helps us understand what happened in the next phase of the transport revolution. The main line of the Peak Forest no longer goes to Buxworth but continues down the arm to Whaley Bridge, where it ends in an extraordinary interchange building. In 1789 Jessop began work on his first major canal scheme, the Cromford Canal, promoted by the wealthy cotton pioneer, Sir Richard Arkwright, to link the town that had grown up around his mill to the Erewash. In 1825 Jessop's son, Josias, was to look for ways to link the cotton areas of Lancashire and Derbyshire, by joining the Cromford and Peak Forest canals. In between lay the hills and valleys of the Derbyshire Peaks, so the answer was not a canal, but the Cromford & High Peak Railway. Although this was to be built on the tramway principle, with mighty steam-engines standing above the steep inclines, the Act mentioned a new element. It spoke of 'Locomotive Steam Engines' as well. For this was the year that saw the opening of the Stockton & Darlington Railway. So, at Whaley Bridge, the canal runs straight into the simple building – and comes to a dead end. Out of the other side, under a wide arch, the rails emerge. This is the railway as the canal's friend – filling in the gaps, joining canals together. Travel the other route down the Peak Forest, however, continue down the Ashton and on into the Rochdale to Castlefield and the junction with that pioneering canal, the Bridgewater, and you can find evidence of a very different story. Here, in the space between the canal and the Irwell, is the old station of the Liverpool & Manchester Railway. This was a route that came into existence only after the most bitter fight with the canal and river navigation proprietors. This railway was no friend of the waterways, but a rival. Just as the Bridgewater Canal had thrust its bold new aqueduct over the old river navigation at Barton, announcing the arrival of a new transport system in the 1760s, so now the viaducts of the railway company proclaimed the new age of 1820s. It is no coincidence that so much transport history is crammed into such a small space. Manchester was 'Cottonopolis', the capital of the textile world – Liverpool its port. The demands of a growing industry constantly created and re-created the elements for a new transport system. The need was met by James Brindley for one generation; by George Stephenson for another. The coming of the railways was to have a profound effect on the canals. Some canal enthusiasts have been heard to curse their arrival. They might take consolation in this thought. If the railways had never arrived, the canals would have been crammed with trading boats – and there would have been no room left for the pleasure-boats of today.

LOST AND . . .

Optimism was the keynote of the mania years. Every canal, it seemed, was destined to be a Bridgewater or a Birmingham, paying handsome dividends into the indefinite future. In the short term, some were highly successful, some considerably less so; in the long term, a few seemed destined to disappear – literally – without trace. The most vulnerable canals were those built to serve a particular industry in a specific region, for if that industry failed, then the canal failed with it. And even if the industry prospered, if a better method of moving goods was found then the canal was equally vulnerable. The Aberdare Canal is a typical example.

The canal was just one of a number built in the valleys of South Wales to serve the burgeoning coal-mining and iron-making region. At its southern end it connected with the Glamorganshire Canal, which then ran on down to Cardiff, the docks and the sea. Like the Brecon & Abergavenny, its connections with the industries of the hills were made through tramways. It almost died before it had a chance to begin, for the canal had scarcely opened when the local iron industry went into decline. Fortunately, a number of collieries were opened in the region, producing that valuable commodity, Welsh steam coal, which was much in demand for industry and, in the nineteenth century, for the rapidly developing railway system. Sadly for the Aberdare Canal, the very rail system that had demanded the coal was to be the system which, in time, was itself to carry the coal. The canal system of South Wales carried within itself the seeds of its own destruction.

The Aberdare Canal closed in 1900 and it takes a certain amount of detective work to find the surviving traces. An obvious starting point is Joseph Priestley's comprehensive list of 'Navigable Rivers and Canals' published in 1831. It is reasonably precise:

> Its course from the Glamorganshire Canal is along the western side of the Cynon Valley, nearly parallel to the river of that name, and having passed Aberrammon it terminates at Ynys Cynon, about three-quarters of a mile from Aberdare.

The Grantham Canal, overgrown and derelict in 1961.

So, all one has to do is hunt around near Aberdare, somewhere on the west side of the valley. But this will not lead one to the scant remains of the canal: the old basin, when finally located, turns up not to the west, but to the east of Aberdare, near the modern sports complex. It is not easy to find, and some may think the effort is not worth making. The old basin was a dead-end, and is now an area of weeds and brackish water, where the stumpy pillar of an old wharf crane is almost the only survivor of the working days of the canal. Following the line of the canal is not a great deal more rewarding. It degenerates into a weedy, overgrown ditch, beside which is a tangle of woodland that droops down, trailing branches into the dark water. Even this rump end of a canal soon comes to a halt, for the old track was bought up and used as the route for a road down the valley. It is still possible – if only just – however, to track the canal down the narrow valley, past the long terraces, picked out in once-bright colours. They cling tenaciously to the hillside that rises above them, its promise of open countryside a little spoiled by the mounds of slag-heaps that line the ridge. Down in the valley, industrial memories of colliery and iron work are being bulldozed away, though there seems no great rush to look for alternatives. At Abercynon, a pub, 'The Navigation', is almost the only sign that a canal ever came this way at all.

Abercynon is the key to the death of the canals of the South Wales valleys. The Aberdare was continued north by tramway, the track of which can be traced from the end of the disused basin. At the Abercynon end, another tramway led from a wharf on the Glamorgan Canal – a site now occupied by the local fire station – to the iron works of Merthyr Tydfil. It was here on 13 February 1804 that the Cornish engineer Richard Trevithick arrived with his small, powerful but primitive steam locomotive to see if it could haul a trainload of waggons. It succeeded, and though there was no great rush to the inventor's door, the future of the steam railway was assured. Traces of the old tramway, with its distinctive rows of stone sleeper-blocks, can still be found, particularly on the section beyond Quakers Yard. The tramway on which this historic experiment was carried out – the first ever steam locomotive to haul a train along iron rails – has survived in rather more recognizable form than many of the canals of the region that the railways were eventually to supplant.

There was never anything romantic about the canals of South Wales. Boats were crude, usually open to the elements and pointed at both ends. Journeys were generally short, so there was no need for a cabin. Cargoes comprised coal, iron and ore, and on every side there was more than ample evidence of where that cargo had its origins. Furnaces spouted flames, the headgear whirred above the coal pits and the spoil heaps rose above the terraces. In spite of the changes brought about in recent years, it is still easy to imagine the scenes of the heyday of the South Wales canals, but it is not

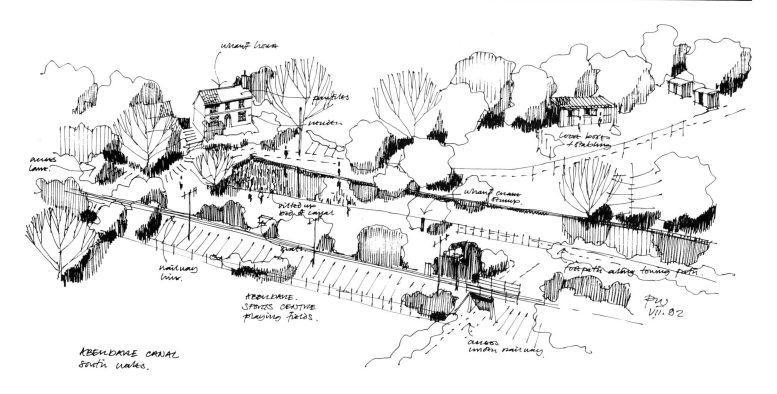

always that simple to understand the origins of the canals that grew up in England in the same period. A few of the canals of 1793, however, still display ample evidence of their origins. This is certainly true of the Nutbrook Canal.

The Nutbrook Canal's task was simple, to take coal from a number of pits in the area of Shipley, in the heart of D.H. Lawrence country near Eastwood, south to the Erewash at Stanton. It was quite a minor affair, a mere four and a half miles long, with a system of railways connecting it to the various mines. These were temporary locations, moved as the colliery sites changed. Today, the colliers of two centuries ago, would be astonished to learn the region has become the Shipley country park, with only the words 'Mines (dis)' on the Ordnance Survey map to speak of those days. The canal has not, however, vanished from the landscape. The reservoirs here and at Mapperley still exist – even if their present claim to fame rests rather more on a record-breaking carp caught here in 1930 than on any industrial connection. And a short section of canal has survived in an area that has been associated with iron-making for centuries, and remains so still. New Stanton is an area with the desolation that so often follows the death of an industrial giant: empty spaces of grimy dust bearing the scars of old rail tracks and sidings. Straggles of terraces are left, somewhat forlornly, staring out over a space that once held the works to which its occupants would have come each day. Through this wasteland the old canal provides a green, or at any

The Aberdare Canal originally led from Aberdare via Mountain Ash to Abercynon, to meet the Glamorganshire Canal, and thence run on to Cardiff. At Aberdare the line of a tramroad that used to connect the town with the canal basin is now an access track. The 6 1/2-mile length has been almost entirely obliterated, part of the towpath now forming a public walkway.
The wharf house and a forlorn old crane base are the sole surviving remnants, the former possessing a certain style – rendered, with arched doorcase and tripartite window casements. The canal itself is silted up and full of aquatic weed. Only 60-foot boats could navigate originally, often with very low headroom at bridges due to mining subsidence.

rate, greenish corridor. A former lock, now a concrete waterfall, acts as a mini-cascade. Coot and moorhen seem to find the place to their liking, swans nest among the reeds, and anglers sit patiently under the obligatory umbrellas. The canal is just about the most attractive thing in the whole area, but now it slides under a bridge, lowered to make life easy for the modern motorist, and comes to a dead halt at a modern factory. The Stanton Ironworks was established here in the 1840s with three blast-furnaces, and the canal ran through the middle. There are solid memories of those days in the hunks of pig-iron still embedded in the canal side. But the modern works has no interest in water transport, and the canal simply comes to a full stop.

The Nutbrook Canal was privately financed, but, as one commentator of the time wrote, 'It is not on that account an undertaking devoid of benefit to the public.' Today, it is derelict, but the same statement still applies. The nearby Derby Canal, although financed in a more conventional manner, fulfils much the same function as the Nutbrook, serving an area of coal mines and industrial concerns. It is a canal whose history is sadly a great deal more interesting than its few remains. The engineer was Benjamin Outram, who was a partner with Jessop in the Butterley Ironworks not far

OPPOSITE: *Patterns of wear: a rope-scored bridge, Kintbury, Kennet & Avon.*

BELOW: *The Nutbrook Canal linked the White House Junction on the Erewash Canal near Sandiacre, passing through the Stanton Ironworks via Kirk Hallam and Shipley collieries, to Shipley wharf - a distance of some four and a half miles. The modern steel pipe works at Stanton have caused the canal to be culverted and reduced to a series of fishing lengths and adjoining pools. Old Furnace Lock and Manners colliery convey the purpose behind its construction - but Peewit Lock and Nutbrook Lock make it sound like something out of a Rupert Bear adventure.*

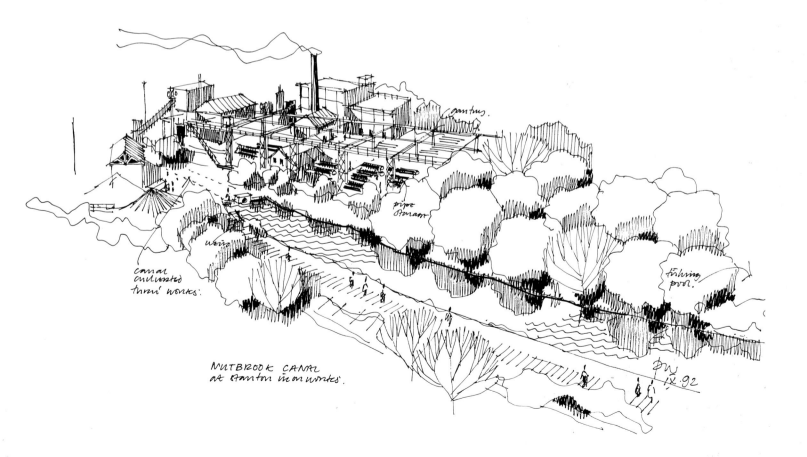

NUTBROOK CANAL
at Stanton iron works.

Narrow-boats line up outside the Stanton Ironworks on the Nutbrook Canal.

from the line of the proposed canal. Outram's route ran from the Trent & Mersey at Swarkestone to the Erewash at Sandiacre, but Jessop suggested a spur to Little Eaton, which could then be extended by a tramway to Denby. This was to be Outram's first tramway – but by no means his last. Indeed, he built so many that some have suggested that the very name was simply an abbreviation of 'Outram way' – a happy thought, but untrue. This line was, in fact, known as 'The Gang Road'. Although it was an early example of a tramway, it was not in itself anything to cause great stir or comment. What makes it more interesting is the fact that it was one of the very first examples of containerization. The tops of the waggons could be lifted off the wheels and dropped neatly into place on a waiting boat. It ran until 1908, long enough to be photographed and the wharf site at Little Eaton is just about recognizable, though the canal has long since been filled in. An old warehouse with a clock tower survives. The assiduous detective may find traces of the tramway in the unlikeliest places. The rails were mounted on roughly triangular-shaped stone blocks. Two holes were drilled, plugged with wood and the rails attached by spikes – a system rather like using a modern rawlplug. A few blocks remain *in situ* – and a great many more can be found in local stone walls.

The fate of the canal's other innovative feature is an even sorrier tale. The canal world of 1793 was a small one, with the same personalities popping up all over the place – and the same was true of the industrial world as a whole. Jessop was a partner of Outram and had Telford as his assistant. Telford worked in Shropshire with the local ironmaster Wilkinson. It was inevitable that these four men, all outstanding in their own ways, should share ideas. So, while Telford and Wilkinson experimented with iron aqueducts at Longdon-upon-Tern as a prelude to the grand theme of Pont Cysyllte, Outram was also designing a cast-iron aqueduct across the Derwent at Derby. It can reasonably be claimed that Outram's was the first – if only because it was finished a few weeks before Longdon's. So where is this great technological innovation? It remained in place until the 1970s, though latterly overshadowed, literally, by a Victorian road bridge. Then along came that inevitable feature of modern city life – the ring road – and down came the aqueduct. It was stored by the Highways Department in 'safe keeping' – safe, that is, until someone got fed up with it littering the depot and sold it for scrap. All that remain are two sections which, to complete this tale of woe, do not even fit together.

The Derby Canal has not, then, been a great success as far as preservation is concerned. It looks promising enough if one goes to Swarkestone, where it leaves the Trent & Mersey. Boats are moored at the wharf and there is a simple wharf cottage. An attractive canal bridge rises above the water, promising delights to come. Sadly, it proves an illusion. The bridge that looks so fine when seen from the front appears as a shambles at the back, and the canal deteriorates into a muddy, reedy swamp. But the towpath at least remains in use as a footpath and cyclepath. The visit to Swarkestone does, however, show a different example of how a canal can be treated. The section of the Trent & Mersey from Horninglow Wharf near Burton-on-Trent to Derwent Mouth was built with broad locks, 13 feet 6 inches wide, and here is one of them. It is a common enough scene – the lock, the black-and-white balance beams, the unpretentious lock cottage. What distinguishes it is the detail, and the use of materials. The red brick of the bridge contrasts with the pale stone of the coping, and stone defines the arch and protects it from damage. The bank too is protected by rough stone setts, which rise up above the towpath, doing a useful job of work and at the same time giving a wonderful texture. Rough stands against smooth; straight line juxtaposes with curve. One sees it time and again along the canal, but when it all comes together as well as it does here, it still seems a small miracle.

The industrial canals of the 1790s, serving specific needs and threading their way through what many people would regard as unattractive, if not downright ugly, scenery, were candidates for closure on a number of counts. If the industries faded, they lost their trade; if the industries were very successful, then there was an incentive to replace the canals with railways; and if trade did fall off, then such canals were unlikely to head any list of candidates for restoration. The last verdict might

The Derby Canal, some 14 1/2 miles long, originally connected the Erewash Canal at Sandiacre to the Trent & Mersey at Swarkestone. Branches ran into the city to Gandy's Wharf and the gasworks, and there was a 3 1/2-mile-long extension to Little Eaton. Today, stretches of the canal bed and towpath form a public walkway amenity and, sadly the canal basin, bridge and cottage at Swarkestone are all that remain to convey a reminder of the once-thriving commercial activity. Swarkestone boat-club moorings occupy the site, and use the wharf and cottage to serve what now operate as 'off-channel' moorings to the Trent & Mersey Canal. Hipped roofs, slate, brick and stone, arched spans and faceted bays are the features that created the architectural personality of the canal.

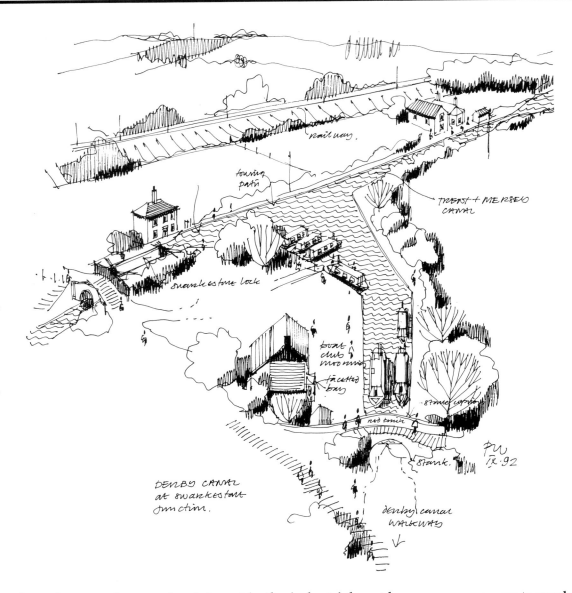

railway.

towing path

TRENT + MERSEY CANAL

swarkestone lock

boat club moorings

faceted bay

stone cyclid

red brick

DERBY CANAL at swarkestone junction.

stank.

PW IX·92

derby canal WALKWAY

OPPOSITE: *Laggan, Caledonian Canal.*

be unjust – and a growing interest in the industrial past has gone some way towards reversing it – but that was the story that unfolded around two canals built to serve the South Yorkshire coal field.

The Dearne & Dove may have a romantic-sounding name, but it was very much the industrial canal. It begins at Swinton, on the River Don section of the Sheffield & South Yorkshire, in a positive flurry of activity and in an area of enormous contrasts. The Don here has a neat, spruce lock with an equally neat, spruce cabin from which craft movement is controlled by traffic lights. The Dearne & Dove begins with a flight of six broad locks – not as big as those on their neighbouring waterway, but still impressive in scale. The walls are made of rough, dressed sandstone blocks, above

The navigation commenced at a junction with the old River Don Navigation (now known as the Sheffield & South Yorkshire Navigation) at Swinton, and proceeded to Barnsley Junction, where it connected with the Barnsley Canal operated by the Aire & Calder Navigation Company. Shown here are part of the six locks at Swinton, the bottom pounds being occupied by Waddingtons, the freight carrier. The somewhat 'picturesque' squalor of this operation conceals an extremely attractive flight of locks, the round pounds of which are being used for boat repair/dry dock purposes. The three pubs – the Towpath, the former Red House and the delightful pile of buildings called the Ship – create an interesting potential conservation area.

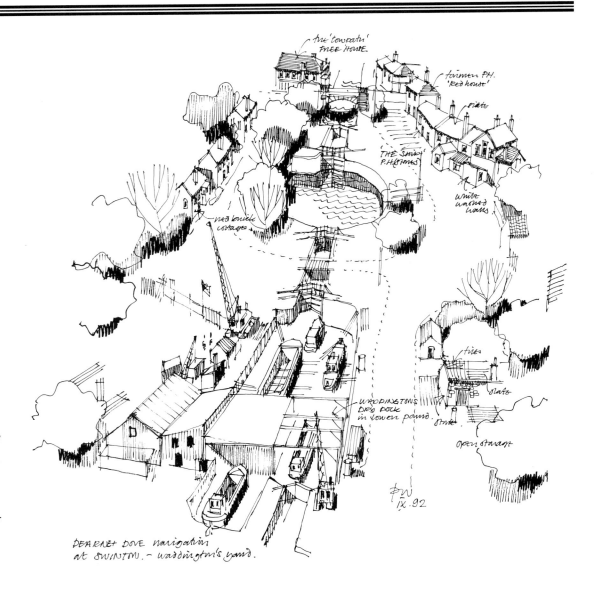

DEARNE+ DOVE navigation at SWINTON. – waddington's yard.

which are the massive balance beams, now slowly rotting and splintering. But this first part of the canal is still used as a basin and dry dock by Waddingtons, who have been running barges from Swinton for more than 200 years. Barges lie around in a seemingly higgledy-piggledy way, but no doubt there is method here somewhere. It is safe to assume that a company that has been handling boats for a couple of centuries knows what it is doing, even if everything looks incomprehensible to the amateur observer. There is a real sense of purpose, and it is not hard to imagine the sweating boatmen manhandling their vessels up the flight in the old days, and then pausing for refreshment, able to choose between the Towpath pub on one side and the Ship on the other. Then comes the disappointment: cross the road, and the canal simply comes to a full stop.

The canal reappears near Wombwell in a man-made landscape of hills, bordered by the old spoil heaps from the local collieries. This was the landscape that wrote the end to the story of the Dearne & Dove. Mining subsidence destroyed the canal, and no one wanted to make it good. Yet something of the nature of the old canal survives. A bridge, built of the same rough-hewn blocks that are such a feature of the locks down by the Don, has a splendid solidity. It looks as if it could stand for ever. The area is overgrown, but at least the canal is still a recognizable feature. The wharf area is almost overwhelmed by the spoil tips, riven by deep gullies, yet it somehow appears to be quite attractive, a touch of greenery in a dead land. On the other side of the bridge, the canal disappears into a reedy wilderness. But at Wombwell itself the canal does again provide relief in an otherwise drab landscape, and the houses have gardens that run down to the water's edge.

Perversely, perhaps, the most interesting sections are not on the main line of the canal, but along its branches. The Worsbrough branch is in quite good order, following an easy course along the Dove valley. At the head of the valley a large reservoir was built, and an agreement was reached that also allowed water to be supplied for the mill

A scene of dereliction at Worsbrough locks on the Dearne & Dove in 1954.

pond of Worsbrough Mill. The mill has survived in somewhat better order than the canal, growing with the years as steam power was added to water power. Steam appeared again, in more dramatic form on the Elsecar branch. Here one can see precisely the job of work that the canal was built to do: carrying coal. The colliery was owned by Earl Fitzwilliam, who in the nineteenth century was able to visit it by rail, stopping at his very own private station. At the time the canal was being built, however, the colliery was acquiring a magnificent new pumping engine. From the outside, one sees the overhead beam poking out through the top of the tall engine house, and pump rods dangling above the shaft. It looks at first glance like the familiar Cornish engine houses, and so in a way it is. But inside the engine is very different. This was a Newcomen-type or atmospheric engine. The top of the cylinder is open. Steam is allowed in, then condensed by spraying with cold water, creating a vacuum. Air pressure then forces the piston down. Once at the bottom of the cylinder, the weight of the rods at the other end of the beam hauls the piston up again, ready for the next cycle. It is a remarkably robust engine, but a great eater-up of fuel – happily for its survival, this was not a problem at a busy coal mine. So often when travelling the canals we lose sight of their original function and drift along in a dreamy, romantic haze. Elsecar restores the balance, bringing us hard up against the noisy, dirty world of mining. Yet it too has its romance – the romance of an age where huge, ponderous machines began to change the face of Britain. It is good news that work is already in hand to bring this branch back to life.

The Dearne & Dove connected directly with another canal of 1793. The Barnsley Canal was to be yet another Jessop project: one can only marvel at the amount of work that he took on at this time. It ran from Barnsley to the Aire & Calder near Wakefield and, like the Derby, has lost its most impressive feature, in this case a five-arched aqueduct across the Dearne valley. Unlike the Derby, however, enough of the canal remains to give a good notion of its character.

The first section near Royston is as unappealing as one can imagine. By the roadside is an enclosure, created out of a ramshackle array of wire and corrugated-iron sheets, behind which dogs bark and geese hiss in a seemingly endless display of bad temper. The canal itself, a rubbish-filled ditch, writhes away through an industrial wasteland dominated by a large drift mine, colliery headstocks and grey spoil heaps. If this were all there was to it, one might be tempted to forget the whole affair and hunt out some more attractive canal to look at. But things begin to perk up at Royston Bridge Wharf. True, the industrial connection is still there, for this was a colliery staithe, but a more surprising rural note is also struck: a house built up against the canal embankment has the word 'Dairy' neatly carved into the lintel of a basement doorway. Further along the route, the Ship Inn stands as a

Washing day, narrow boat Warbler.

Part of the important Aire & Calder Navigation system, the Barnsley Canal – some 14 miles long – was intimately connected with the coal-mining industry. Mines at New Sharlston, Ryhill Main, Hodroyd colliery, Wood Moor, Carlton Main, Monckton Main and New Gawber all had their own wharves and staithes. Some later became modern drift-mines, like the one shown near Royston, the former Hodroyd colliery, and now face closure or an uncertain future. The traffic has long left the canal, and pubs like the Ship and the red-brick corner shops stand huddled against the dust and muck from passing lorries. However, demolition has created the odd pocket of open space for walks, but the canal bed itself looks unsightly and dangerous.

marker for the line of the old canal, though one might not guess it from the sign: a fully rigged ship appears to be beating its way round Cape Horn, not quite what one expects from a barge on its way from Barnsley to Wakefield.

At Old Royston the whole character of the canal changes. Here, amid a countryside dotted by mining towns and villages, is a deep, wooded cutting as peaceful and remote as Tring on the Grand Union, or one of the artificial canyons of the Shropshire Union. It is crossed by a road bridge, which utilizes all kinds of materials – iron girders, blue engineering bricks and stone blocks. But it is the cutting itself that is the real focus of attention, for here one can still get a glimpse of the effort required to force a route through a rugged landscape. Great blocks of sandstone litter the slopes, blown out by the canal-builders, and left where they lay. The loose spoil was taken out of the deep, dark cutting and used to build up the embankment across the next valley. It was here that Jessop built the reservoirs which were to ensure that the canal, though its traffic might dry up, would never itself dry up for lack of water. He built a large reservoir at Wintersett, and a second was later added at Cold Hiendly.

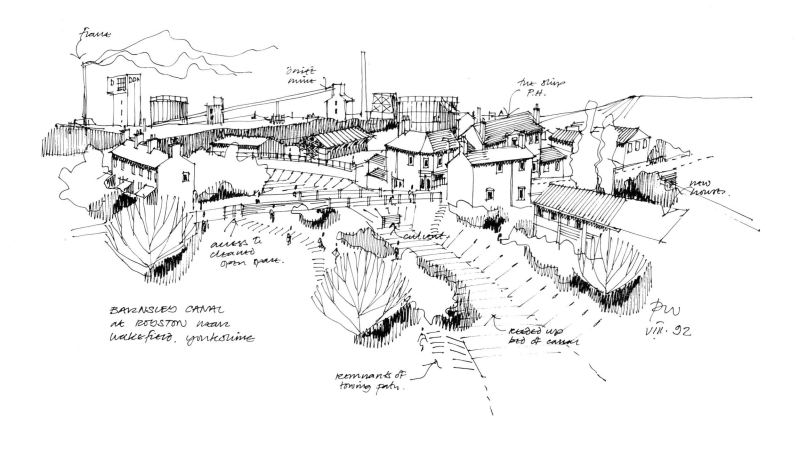

BARNSLEY CANAL at ROYSTON near wakefield. yorkshire

Traffic has long since ceased on the Barnsley, but at least it ends at a working waterway, the Wakefield branch of the Aire & Calder, where a large power station shows where modern canal trade has its best hope of surviving. It was the Duke of Bridgewater who said that the successful canal had 'coal at the heel of it', and the statement remains true today. Keeping a power station supplied is still a job that can be performed efficiently and economically by water. The equation, however, requires both supplier (the colliery) and customer (the power station) to have waterside sites. If any of the variables are changed, then the traffic is likely to fail. So canals such as the Barnsley fell into dereliction – but at least it is easy to see why they were built, and why they prospered. With other canals the task is by no means as simple. Why, for example, did it ever seem a good idea to build the Caistor Canal? Even allowing for the rush of blood to the speculative brain that marked out 1793 as a special year, it is difficult to see what trade – actual, potential or even fanciful – could have persuaded investors to put their cash into the scheme. It was certainly not a canal that met the Duke of Bridgewater's criterion – it did not have coal at the heel. It had little of anything at the heel.

Warehouses and a small hand-crane by Barnsley basin.

A busy scene at the Barnsley Canal staithes at Edmund Main Colliery, 1859.

The Caistor was, at its best, a puny affair. The idea was to create a waterway that would run from the New River Ancholme to Caistor, a modest enough ambition, one would have thought. It was to be a distance of little more than seven miles, over the conveniently flat land of Lincolnshire – but the canal never made it. It struggled on as far as Moortown, halfway along the route, and there it rested, never to move again. If ever a canal could be said to join nothing to nowhere, this is it. There are no towns of any size en route, nothing to create traffic other than what contemporaries called 'the surplus agricultural produce of the north of Lincolnshire'. It comes as no surprise to find that this canal has left little of interest behind – a mile or so of reedy water near the junction, a flattened area near South Kelsey that was once a wharf, and that is about it. You get a notion of what it was like from the Ancholme, running as straight as a Dutch dike through the flat lands, where the only notable features occur where waterway and road enjoy one of their rare meetings. At Brandy Wharf, appropriately given the name, the solitary warehouse has been converted into a pub. There were attempts to complete the Caistor, and even to extend it as far as Market Rasen at the foot of the Wolds, but nothing ever came of them. Jessop was asked to report on the canal's feasibility in the planning stage in 1792. This he duly did but then, sensible man, he pocketed his fee and had nothing more to do with it.

Canals through agricultural areas, which failed to arrive at any industrial regions or to meet any substantial towns, were rarely a success. Work had begun in 1791 to extend navigation from Leicester via the Wreake & Eye to Melton Mowbray. In 1793 enthusiasts put forward plans for another extension – on by canal from Melton Mowbray to Oakham. The Oakham Canal meandered in a somewhat meaningless fashion, as vague in its sense of direction as anything contrived by Mr Brindley at his most perverse – starting off by heading due north and ending up heading west. In between was a long detour round Stapleford Park. Lord Harborough was no lover of innovation, and the park was to be the scene of a notable battle, when railway survey-ors tried to find their own route through the region. The canal company, it seems, accepted the inevitable and made a detour, leaving his lordship undisturbed and the countryside at peace – and so too did the railway on what was known as 'Lord Harborough's Curve'. The railways' battle with his lordship was, literally, a bloody affair, a battle between navvies and gamekeepers; the battle between railway and canal company was a good deal more peaceful, but, at the same time, a good deal more decisive. The canal seemed to offer, for part of the way at least, the best route through the undulating countryside. So the railway bought it up, and began building its tracks along the canal banks. The result was inevitable, but good news for the shareholders: the amount paid in compensation was considerably more than it ever seemed likely the canal would earn from traffic and trade. So it was that the Oakham

Narrow boat tensioning chains.

The former Oakham Canal
Company offices and wharf at
Market Overton are the best
surviving remnants of this
agricultural canal, which never
really paid its way. Money was
not spared, however, on the
wharf toll-house and warehouses.
Cream limestone with slate
roofs here form a delightful
roadside complex. Tudor-style
arches, complete with drip
mouldings, demonstrate an
almost 'collegiate' style and
authority. Re-roofed in modern
tiles, the simple canalside cottages have been much altered –
each with its new attached
garage block. The bridge has
gone, to level the road, but
water can still be seen on the
side opposite the wharf. Today
the peace is shattered by
Tornado aircraft
operating from
the nearby RAF
station.

Canal slipped into oblivion so completely that for many miles a keen eye or sound local knowledge have to be relied upon to uncover it.

At first, near Oakham itself, the canal is plain enough, if uninspiring – the muddy, reedy ditch so characteristic of the abandoned waterway. Features that might have provided a focus of interest have gone. Bridges that must once have been graceful arches have been dropped, reduced to graceless slabs standing only inches above the brackish water. There is more romance on the nearby railway, which has at least preserved an old-fashioned signal box by the level-crossing, yet there are still reminders of better days. At The Wharf, near Market Overton, stand wharf cottages, built of brick and stone. Now they are attractive country houses, but one can still detect the regular pattern of small windows of what was once a substantial warehouse. The canal itself has all but disappeared, although sometimes quite surprising clues appear to indicate its presence. Near Edmondthorpe, an old mile-post stands in a garden. Once it passed on the news to boatmen that they were eight miles from Melton Mowbray and seven and a half from Oakham. But it is

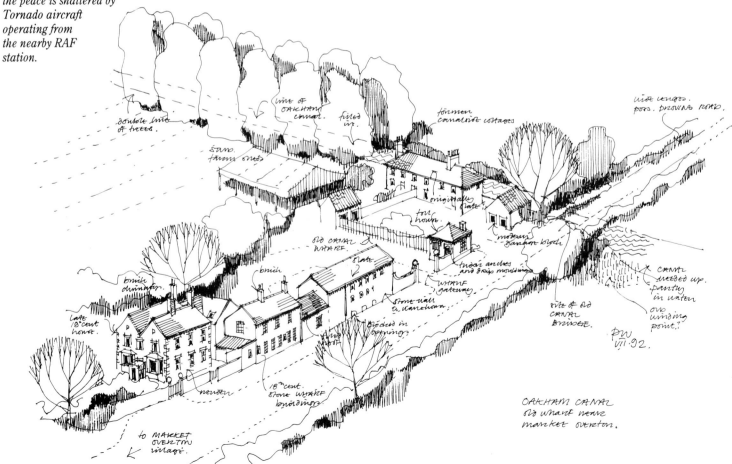

OAKHAM CANAL
old wharf near
market overton.

soon easy to see why the information became useless, as the railway embankment cuts ruthlessly across the line of the old canal. The engineers were not, in any case, probably feeling any too kind-hearted, as they had just been forced to make that infamous detour. Stapleford Park remains untouched, defiantly picturesque, with its little *cottages ornées* unsullied by canal or rail.

The Oakham is little more than a shadow on the land, but a nearby contemporary has fared a good deal better. The Grantham Canal was promoted to link that town to the waterway system, to avoid the heavy expenses of hauling freight overland from Newark on the River Trent or from Boston. The locals turned, like everyone else at that time, to Jessop, who proposed a canal from Radcliffe in 1792. That was rejected, but the following year he tried again with a new connection to the Trent at Nottingham. This was approved and work on the 33-mile-long waterway began. It was to be a splendid affair, with 14-foot wide locks, able to take a single barge or a pair of narrow-boats. Jessop saw no need for any very grand earthworks, and for once he was content to let the canal meander along to find its level through the Vale of Belvoir. One distinctive feature, which seems obvious now but was quite new at the time, was the use of flood-water reservoirs. Jessop realized that by not taking river water he could placate the mill owners, and wrote a treatise on the subject.

The canal prospered, but the day inevitably arrived when a railway was opened from Grantham to Nottingham. The railway went through complicated – and expensive – birth-pangs, so expensive indeed that the company found itself too short of cash to pay the Canal Company the compensation it had agreed to when the Act of Parliament was granted. That was the start of the canal's woes, though the years were not all gloomy: it even fulfilled a grandly patriotic role in the First World War, when it was used to move men and stores of the Machine Gun Corps. But declining trade meant less money for maintenance; and poor maintenance meant a further decline in trade. In 1924 traffic came to a halt and in 1936 the canal was abandoned. And that could well have been that – the Grantham could have gone the way of the Oakham. But canals do more than just provide a pathway for boats: they carry another valuable commodity, water. Local farmers valued the canal, even if traders did not, so the Grantham was kept in water and is still in astonishingly good condition. People looking at its neat towpaths and clear, deep water might well think that it is in better condition than many canals allegedly open for boating.

The canal is such an obvious delight that the wonder is that it was not restored for pleasure-boating years ago. One major obstacle was the embankment that replaced a bridge, which carried a railway to the local ironstone mines. This was a double indignity for the canal to bear. In its working days, it would have seemed ideally suited for carrying the ore in the wide barges, but instead the railway was built alongside and over the waterway, taking away the trade. Then when the Great Northern Railway, which owned both railway and canal, decided to replace the bridge, they dumped their bank slap-bang in the middle of the canal, and reduced the waterway to a culvert. The GNR never made any secret of its views of all canals as anachronistic nuisances. The line has been closed for some time, but it is only recently that permission has been granted to open it up to enable the canal to come through.

There is no shortage of problems facing the restorers. Locks have been cascaded, a few old hump-backed bridges removed and replaced by horizontal spans sited low over the water and, more seriously, the connection with the Trent has been lost. The Grantham end has also been built over, but this is no great handicap – it simply makes the route a little shorter; but to be viable, the canal needs to have access to the rest of the waterway system. This should not, in fact, prove too difficult, for it should be possible to gain access to the river via the new marina. The one thing that proves abundantly clear to anyone visiting the Grantham is that the effort of restoration is well worth while. There are 18 locks on the canal, but between Fosse Top at the Nottingham end and the bottom of the Woolsthorpe flight there is a 20-mile pound. It wanders through the lovely Vale of Belvoir, still dominated by its ancient castle. The scenery is splendid, and so too is the canal itself.

The Nottingham end brings reminders that, however lovely the canal might be, it was built for work. At Cotgrave, the locks carry the canal up past the colliery, the showpiece of the Coal Board's modernization programme back in the 1950s. But if coal-mining is common, gypsum-mining was in progress when the canal was built.

The canal at Grantham in its working days.

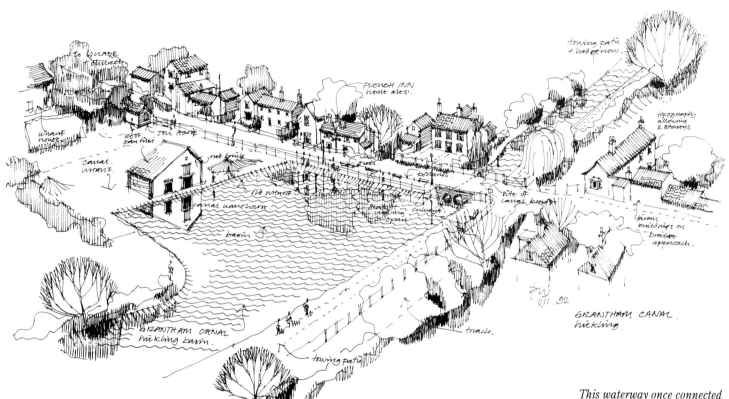

Gypsum may be a very useful substance – it forms the basis of plaster of Paris – and would have been a welcome load for the boats, but it was a confounded nuisance for the engineers. The canal showed an unhappy tendency to leak wherever it passed through the mineral strata. As this area is left behind, so the canal takes on its more typical rural air. Little villages along the way were served by wharves or small basins. The first task of the Grantham Canal Restoration Society was to clear the basin at Hickling. It is an area of simple, but considerable charm: a few houses, a pub, a little red-brick wharf building and a broad sheet of water, now well populated by ducks, who know a good thing when it is presented to them. People are now no more to them than perambulating bread bags. Someone took the decision to reinforce the boating image by installing a row of small bollards on the pavement by the wharf. A good idea, but sadly they were set on the road side, not the water side. Handy for tying up cars, but any boats using them would'have to stretch lines across the pavement. It is the kind of absurdity that is always liable to occur when designers try to 'give character' to an area by cosmetic change, rather than thinking of function first. One should not be too unkind, however: the bollards are at least a sign that the basin is now regarded as a valuable amenity.

This waterway once connected to the River Trent in Nottingham and proceeded via Cropwell, Redmile and Woolsthorpe to Grantham in Lincolnshire - a distance of 33 miles. At several small agricultural villages, barns and wharves enabled coal and produce to be unloaded and stored. It was a wide waterway, passing boats up to 75 feet long, but was never that commercially successful. At Hickling a fine basin still exists, together with a red-brick warehouse under a pantiled roof. The bridge has been demolished and replaced by culverts, and an odd canal 'viewing area' – complete with seats and bollards on the roadside (so that cars can be tied up?) – has been introduced. The Plough Inn is nearby, to drown one's sorrows!

There is an excellent example of just how attractive buildings can be, even when they are wholly functional, at Harby. They come as something of a surprise: one somehow does not expect to find working maintenance buildings on a derelict canal, but at least they explain why so much of the canal is in good condition. The Grantham is full of interest. Near Bottesford is Muston Gorse Wharf – no ordinary wharf, but one built by the Duke of Rutland, from which a private tramway ran up the hill to take supplies to Belvoir Castle. Nearby, but far less obvious, is a feeder from the reservoir at Knipton, out of sight behind Blackberry hill, on which the castle stands. One advantage of being a Duke is that you can have your very own tramway clanking along wherever you like, but you can also insist that a canal company keeps its feeder out of view. It runs through the hill in a tunnel – which must have been a considerable expense to build and a considerable nuisance to maintain. Now the canal enters one of its most attractive reaches. The countryside is gentle; the castle on the hill dramatic; and the bridge at Stenwith crosses the canal as a beautifully welcoming arch.

Beyond the bridge a broad, grassy track is not an unusually grand towpath, but the track of the old ironstone railway. The old boatmen, who must have fumed at seeing 'their' cargo being hauled along beside them, might have got some satisfaction from knowing that boats would still be plying the waterway long after the track had been lifted. The canal got the last laugh after all, for this was the first section on which the Canal Society was able to establish a trip-boat. This stretch leads up to the flight of locks and the pub officially known as the Rutland Arms, unofficially as 'The Dirty Duck'. In this final section of the canal, the problems of the landscape, which had been evaded for so long, finally had to be faced. The eight locks lift it up to the level of the last pound, leaving one hill to be pierced in the long and attractive Harlaxton cutting.

Walkers can enjoy the Grantham Canal, for the towpath is open. Sooner or later boats will surely be using it again, and it seems certain to prove a very popular route. Some will also find a certain ironic pleasure in travelling on a nationalized transport route to the home town of the arch-priestess of private enterprise. But if the Grantham Canal seems likely to enjoy a prosperous future, other disused canals seem likely to remain in neglect. Some vanished even before the railway age got under way.

Cornwall got its first canal when work began in 1773 on the St Columb Canal near Newquay. Its promoter, John Edyvean, was a man of great ideas. He proposed a canal that would link the north coast at Bude with the Tamar at Calstock, and followed that up with a plan for a canal that would run the entire length of the country. These schemes came to nothing, but the St Columb was begun, if not finished. It was, like the later Bude Canal, designed to take sand for fertilizer to use on inland farms. There were two sections, one from Trenance Point towards St Columb Major, the other from Lusty Glaze Bay in Newquay, which headed inland but never reached any particular destina-

A candidate for restoration: pleasure-steamers at Craigmarloch on the Forth & Clyde in happier days.

tion. As early as 1824 it was noted that the canals had all but vanished from view, and today just about the only visible sign is a cut in the cliffs at Lusty Glaze up which the tubs of sand were hauled. Assiduous hunters after old canals can also search for the ruined locks of the Looe Union, which ran from Liskeard to Looe. It was while inspecting one of these locks at St Keyne that I was startled to hear the distinctive sound of a fairground organ – and discovered, and delighted in, the Paul Corin Collection of Mechanical Music. Hunting out old canals can have some unexpected bonuses.

At the opposite end of the country, Scotland has some magnificent waterways which deserve to be better known, and some which are now beyond knowing. The Forth & Clyde, scene of John Smeaton's memorably acid exchange with James Brindley, has already been mentioned and is still very much a part of the landscape. The same, alas, cannot be said of one of the earliest and, in some ways, one of the most interesting. The Monkland Canal was designed to join the coal district around Airdrie to Glasgow. James Watt surveyed the route in 1767, but by 1773 only a few miles had been completed. It was only when the Stirling brothers, who had extensive commercial interests in the region, took over the management that things began to move again, and in 1790 work once more got under way. The canal was not at first a great success, but a booming iron trade saw four new works open around Old and New Monkland, and the canal prospered to such an extent that it could no longer cope with traffic at the Blackhill locks on the outskirts of Glasgow. Rather than duplicate the locks, an incline was built in 1850–1, using 'Gazzoons' – Glaswegian caissons.

At its peak, it was taking 60 boats a day up the hill. But then traffic declined, and for a while the Monkland survived largely as a feeder for the Forth & Clyde, but in the 1930s it was abandoned, gazzoons and all, filled in and lost for ever.

Other canals have had equally unhappy fates. Little remains of the Glasgow, Paisley & Ardrossan. The Dingwall Canal, which is really the River Peffray with the kinks removed, is in rather better condition, but the Aberdeenshire is now little more than the remains of a water-supply system to paper mills at Inverurie. Happily, not every closed or abandoned canal has necessarily gone for ever. Work began in 1817 on the Edinburgh & Glasgow Union Canal, which linked the Forth & Clyde to Edinburgh. Its main claim to fame lies in its aqueducts. Any one of them would be sufficient to make the canal memorable, but here there are three. They are similar to Chirk, in being built of stone with an iron trough let into the masonry. The largest strides high above the Avon on 12 arches; the Almond is almost equally grand; but the third is the most unusual, crossing suburban rooftops as well as the Water of Leith on the outskirts of Edinburgh. Sadly, this beautiful canal has been chopped off at both ends – it no longer unites anything, severed from the Forth & Clyde, and cut off from the terminal basin at Edinburgh. Nevertheless there is a very active restoration programme in hand, and few canals are more deserving of rescue.

Between the extremes of Scotland and Cornwall there is a whole range of canals of every kind and description, which once carried the trade of England. Pride of place should perhaps go to the Sankey Navigation, which can make a claim – depending on how one defines the terms – to being the country's first true canal. The reason that the distinction is generally reserved for the Bridgewater is in part the result of a successful deception. The plan called for a waterway to link St Helens and its surrounding coalfield with the River Mersey. As the Duke of Bridgewater was to discover, there were many vested interests prepared to raise a clamour against cutting wholly artificial waterways across the countryside. Landowners envisaged legions of scruffy boatmen passing their property, creating heaven knows what havoc. Trustees of turnpike trusts saw goods being moved away from land to water; while existing river authorities feared serious inroads into their trade. Then there were those who simply disliked innovation of any kind. River improvement, however, was quite a different matter, a well-established practice. So the promoters pretended that all they proposed was to make the Sankey Brook navigable. The fact that in its upper reaches it was incapable of supporting anything much larger than a paper boat was conveniently ignored. There was a little traffic at the Mersey end, so it seemed quite reasonable to extend that traffic a bit further. So the 'Act for making navigable the River or Brook called Sankey Brook' was duly passed in 1755, without a hint of opposition. It was a river navigation in so far as it followed the course of a natural stream, but in all other respects it was a true canal.

Traditional narrow boat painting.

The Sankey slipped into the waterways system all but unnoticed, its significance well hidden. No one came to stare at its wonders or marvel at its trade. It simply did its job, carrying coal and earning a decent profit for its builders. It was not seen as a precedent; no one thought of it as inaugurating a new age of canals. Over the years branches were built, and improvements made, notably an extension to a more suitable Merseyside port at Widnes. The canal did have its moments of excitement, though. It was a somewhat conservatively run affair on the whole. The flats that came up from the Mersey were either hauled along by gangs of men or rowed, but in 1797 the local papers reported a new and exciting development. A vessel loaded with copper slag made a journey down the canal 'by the application of steam only'. This appears to have been a paddle-steamer, worked by a beam engine – but a cumbersone and inefficient one. Work was returned to the gangs of bow-haulers.

Steam did, however, arrive over the canal in the shape of the nine-arched viaduct that carried the Liverpool & Manchester Railway. A print of the time shows what appears to be one of Stephenson's 'Planet' class of locomotives, cheerfully puffing along with its train, while a flat drifts lazily along underneath. The Sankey was actually faced by a much more direct threat to its trade when work began on the St Helens & Runcorn Gap Railway. It was a decidedly odd affair: it hopped over the Liverpool & Manchester Railway via a pair of inclined planes and crossed the Sankey on swing bridges. Drivers had to stop their trains, get out and close the bridges, and open them again for boat traffic when they had got across. It was, not surprisingly, unpopular. In the end the Sankey Canal and the railway joined forces as the St Helens Canal & Railway Company. And for once it was the Canal Company that was in the better condition. The Sankey is still there, but parts have dried up, been built over or simply fallen into ruin – a sad fate to overtake such a pioneering venture.

The Sankey is just the type of canal that one would expect to be overtaken by events. It was built to serve heavy industry and worked its way through a highly industrialized area. It had no through traffic, for it simply ran from the Mersey to St Helens and came to a stop; it supplied a particular need and, as long as the need existed, it remained in profit. At least that need was clearly identified. This was by no means always the case. The Dorset & Somerset Canal was to link the Dorset Stour to the Kennet & Avon Canal, which looked more sensible in 1796, at the tail end of the mania years, than it does today. The Somerset coalfield was still busy; the West of England woollen industry prospered. Curiously, work began on a branch in the middle of the route – and everything was to be done in the very best style: including the use of boat lifts instead of the conventional locks. James Fussell built a lift that worked very well, and would have served its purpose – if the canal had ever been built. Yet surprisingly, perhaps, quite a lot remains for those with a well-developed detective instinct to hunt out. The site of the

Two pioneers: a Mersey flat on the Sankey Canal heads towards the tall viaduct of the Liverpool & Manchester Railway.

trial lift is there, near the crumbling remains of Fussells Iron Works, and Frome locals will be able to point you towards the Roman Wall, actually a masonry embankment.

Less likely still than a Dorset & Somerset Canal is the notion of a canal at Glastonbury. A contemporary print of the opening shows a procession hauled by a spritsail barge, with a military band bringing up the rear, all heading towards the distinctive hill of Glastonbury Tor. From the list of tolls, it looks as though the main commodities that the company expected to carry were coal and cheese. The waterway itself linked Glastonbury to the Bristol Channel, but although it incorporated all the latest technology, including an iron aqueduct over the River Brue, it failed to survive for more than a few years. Begun in 1827, it was just 30 years old when it was bought up, so that the land could be used for railway building, Soon the broad-gauge locomotives of the Somerset Central Railway were pounding along where boats had quietly glided. The lock-keeper's cottage near the Brue aqueduct simply became the crossing-keeper's cottage. Now the railway has disappeared almost as completely as the canal.

Similar stories can be told of canals all over the country – canals which, it seems, are lost for ever, yet one can never be too sure. If one had surveyed the canal scene as recently as 20 years ago, one would have looked at the list of derelict canals and decided that many were destined to stay derelict for ever. Yet many on that list have now either been restored or are being actively worked on, and new schemes are being started all the time. Perhaps this chapter should have been headed 'Lost – but not necessarily gone for ever'. Who knows which ones might find a home among the waterways of the next chapter at some time in the future?

The phrase 'chequered career' could well have been invented with the Stratford-upon-Avon Canal in mind. It began, as did the other 1793 schemes, in a spirit of great optimism. It seemed a sensible, useful route, connecting the navigable River Avon to the recently authorized Worcester & Birmingham Canal. And the proprietors later realized that another useful connection could be made to the Warwick & Birmingham, where the two canals approached each other at Kingswood. The idea of turning Stratford into an inland port was not a new one. In 1677 Andrew Yarranton, an early advocate of improved river transport, looked to Holland and its successful waterway system and dreamed of establishing new centres at Stratford. There was to be a New Brunswick for brewing beer and a New Haarlem for linen manufacture. Nothing happened to that plan, but a century later promoters began to dream of the prosperous port of Stratford. Talk started on the plans in 1775 and, with a rate of progress that was to prove all too common, came to fruition only in 1793. But where Yarranton saw the establishment of industries as the key to the port's success, those who followed paid scant attention to where the trade would be found. None the less, work soon got under way under the expert supervision of William Clowes. He began at the northern end and tried out the latest device to reach the canal-building world, a horse-drawn cutter. It was a flop, so it was back to the familiar scene of navvy groups hacking a way through the land with pick and shovel.

At first everything seemed to go well, and within a year the canal had reached Hockley Heath, a distance of nearly ten miles. The trouble was that the canal was supposed to be 25 1/2 miles long and already all the money had been spent. Back to Parliament the company was forced to go to get authorization to raise more cash. Work began again in 1800 under a new engineer, Samuel Clowes's former assistant, Samuel Porter. He was able to take the canal to Kingswood and the junction with the Warwick & Birmingham, so that now the Stratford was no longer left in unsplendid isolation but had connections at both ends. This did not mean, however, that there

The Ashton Canal – scene of one of the earliest mass attacks by amateur navvies in a great clean-up campaign.

was money left to finish the job, and again the Stratford languished. It took one of the country's most energetic promoters of new transport routes, William James, to get things moving again. James was to move on from canal promotion to railway promotion, but sadly his enthusiasm for his favourite subject led him to neglect his own business interests, and he ended poor and bitter – seeing the schemes he promoted prosper, as his own fortune dwindled and disappeared. That, however, lay in the future; for now he was able to galvanize the seemingly moribund canal scheme. Work re-started in 1813 and, at last, in 1816 the canal was opened.

There was now a through route from the Severn, down the Avon and on up to Birmingham. Unfortunately for the Stratford Canal, it was not the first to offer a good through route between the Severn and Birmingham. The Worcester & Birmingham Canal had had its share of problems but at least there was the excuse of major engineering problems. In its 30-mile route there are five tunnels, one of which at King's Norton is 2,726 yards long, and there are 58 locks, including the longest flight in Britain at Tardebigge. There are 30 locks in the flight, and the boater approaching them is at least spared the daunting sight of the whole lot spread out at once, for they are built on a long, sweeping curve. In spite of the difficulties, the Worcester & Birmingham opened six months before the Stratford – six months which they used to good advantage to grab as much of the trade off the Severn as possible.

It was soon obvious that there was not going to be very much trade on the Southern part of the Stratford Canal, and if there were to be any profits at all they would come at the northern end. But to secure these the company had to offer traders bargain-basement rates, tolls that were far lower than those of their competitors. The company thus began its commercial life with next to no trade on one half of the canal and cheap-rate trade on the other – not an obvious formula for success. When the railways came to the region, the Stratford was at the office door, cap in hand, begging to be taken over. In 1856, the Oxford, Worcester & Wolverhampton Railway duly obliged and soon the canal was part – a rather insignificant part – of the Great Western empire. It struggled on, maintenance neglected, traffic thin and prospects poor, through nationalization until, in 1958, Warwickshire County Council applied for permission to abandon it altogether. This, which could so easily have been the end of the story, turned out to be the beginning of a new tale.

A new movement was beginning to spread through Britain in the 1950s. Enthusiasts were getting together with the idea of preserving, not destroying, the canal system. One activity was 'campaign cruising'. One of the founding-fathers of the movement was L.T.C. Rolt, whose book *Narrow-Boat*, describing a canal trip around England, did much to arouse enthusiasm for the subject. When news came through that all traffic had come to a halt on the Stratford, because a supposedly moveable

L.T.C. Rolt on Cressy; *their journeys together were described in the classic book* Narrow-Boat.

bridge had become a seemingly immoveable obstruction, Rolt informed the authorities of his intention of bringing his now-famous narrow-boat *Cressy* down the canal. Fine, said the authorities, we will move the bridge for you, and so they did, jacking it up to allow *Cressy* through. It was a valuable exercise in canal preservation, and equally valuable for its publicity value. But it was obvious to many people that the Stratford's condition, particularly at the southern end, was bad, getting worse and, if nothing were done, that it would simply cease to exist as a navigable waterway.

A preservation society was formed and had the common sense to do the one thing that saved the canal from closure. It obtained a licence to take a canoe along the canal, and in six weekends early in 1957 a lone, but intrepid canoeist lugged his craft round unworkable locks, fought his way through reed and weed, and completed the journey from Stratford to Hockley Heath. Then he got another licence and fought his way back again. So when the County Council declared its legal right to close the canal, on the grounds that it had remained unused for the statutory three years, it found itself confronted with the irrefutable evidence that this was not so. The canoe had saved the day.

It is one thing to save a canal from closure; quite another to restore it to its former glory. The Canal Society joined forces with the Inland Waterways Association (IWA) and the Coventry Canal Society, and together they approached the National

Trust, which was persuaded, one suspects rather to everyone's surprise, to add the southern section of the Stratford to its collection of stately homes and manor houses. Now work could begin, and the task of seeing it through went to David Hutchings. In a way it was like a replay of the 1800s, when William James bustled in to bring the canal to life. Now it was the turn of Hutchings: a man who did not so much cut red tape as slash it into confetti; a man who set himself a goal and used any and every means to achieve it. His workforce consisted at various times of volunteers, Boy Scouts, soldiers and prisoners from nearby Winson Green. On the whole he seemed to like the prisoners best – partly because any who did not measure up to his work standards were sent packing back to their cells; partly because those who did stay the course worked with real enthusiasm. David Hutchings was not a man for the niceties. He saw the job as essentially simple: the canal was not usable; he would make it usable. From 1961 to 1964 every lock gate but one was replaced, lock chambers rebuilt and most of the canal cleared of weed, silt and debris. At the end it was officially declared navigable, and a grand opening ceremony was performed on 11 July 1964 by the Queen Mother. The Stratford had been saved – and if the Stratford could be restored, so could other canals.

It might seem that with such a history of misfortune behind it, the Stratford Canal should simply have been left to decay into decent obscurity. But the very reasons that the canal was unprofitable in its cargo-carrying days are those that help to make it a success with holiday-makers. The rural character of the country through which it passes, which was the despair of traffic managers desperately hunting for cargo, enables holiday brochures to be filled with phrases such as 'unspoiled charm' and 'peace of the countryside'. At a period when any form of canal travel for pleasure was still a comparative rarity, this was perhaps the only sort of restoration scheme that would have received wholehearted support. Travelling the canal today, its charms are still evident, and it emerges as a canal full of character and interest.

Starting (as the builders did) at the northern end, one might expect, from the map, to find a canal buried among the drearier suburbs of Birmingham. In part it is, but the actual junction with the Worcester & Birmingham at King's Norton turns out to have quite a rural air, with a view of green fields beyond the houses. The main interest, however, lies in the canal structures. The junction house has something of the ponderously pompous about it. A hipped roof rises above a parapet, a classical porch leads to an imposing front door of hexagonal panels – all very pleasant – but the stones of the window surrounds, which juxtapose with the dull red brick of the walls, provide contrast but rather overwhelm the façade. Rather than reflecting the Georgian style of so many canal buildings, it seems to look forward instead to the Victorian age. The junction bridge is a different matter altogether:

Lock gates, Sharpness.

The guillotine lock on the Stratford Canal near King's Norton junction.

every line is a line of grace. The parapets trace great curves as they adapt to take the towpaths of two canals, and the arch abutments are curved to ease the passage of tow ropes round the bend. This same sense of style allied to use reappears in simple details, like the modern moorings, where shapely iron bollards are given a surrounding of stone setts – smooth iron contrasting happily with rough stone.

Beyond the junction, dreariness does set in. Housing estates and minor industries bring little cheer, and the general appearance is not helped by graffiti that might bring a blush even to the cheeks of a Sun reporter. The best-known feature of the area soon appears, the guillotine lock, so called because the gates drop vertically like the lethal blade. It was put there at the insistence of the Worcester & Birmingham to ensure that none of their water slipped away to the Stratford. It is an interesting and unusual feature, but marred by a pipe that crosses the canal at this point, and it is many years since it was last used. There is a short passage through an industrial area, where a factory is reached via a swing bridge, before the canal disappears into the 352-yard-long Brandwood tunnel. In the days of horse-drawn boats, boatmen moved their craft along by pulling on hand rails. This being the Stratford Canal, it was inevitable that Shakespeare would appear somewhere along the way, and here he is peering down from the tunnel portals. Unfortunately, there seems to be no appropriate quotation from the bard – the nearest he got to mentioning waterways was the well-known description of Cleopatra's meeting with Mark Antony on the River Cydnus:

> The barge she sat in, like a burnish'd throne,
> Burn'd on the water.

Nothing that passed down the Stratford Canal was likely to meet those events of splendour, nor be graced with a poop of beaten gold, purple sails or silver oars. So the bard remains speechless as he stares down the length of the cutting. Although the canal is still in the suburbs, it contrives to create its own little self-contained world, a green finger poking out from the city. It can easily be missed altogether by road users: a tall arch across the road would probably be taken for a railway bridge, but is in fact a short aqueduct. A new lift bridge is rather more obvious, and has resulted in a name change for the local pub. The Boatman's Rest is now the Drawbridge, as it was many years before.

There was originally no special provision made for water supply, but the Act of 1815, revising the line, allowed reservoirs to be built at Earlswood. They have

A lifting bridge on the northern Stratford. There is now a new bridge – and a new Drawbridge pub.

become a popular local amenity, known as Earlswood Lakes, used for angling and sailing, and grand enough to have their very own Lake Station on the Birmingham to Stratford Railway. This is an area where the canal seems to withdraw into itself, hiding away in the deep cutting of Salter Street. There are memories of busier times – when there was a lime kiln in Limekiln Lane and the local pub was a brewhouse. But the seclusion has been disturbed in recent years by the arrival of the M42. The older roads of the area, however, provide a convenient way of viewing the canal engineering. They can dip and rise with impunity, taking the straightest and most convenient route, whereas the canal can be seen as a succession of small variations: a slight bank here, a shallow cutting there, or a detour from the straight and narrow to keep a level – small changes that are scarcely noted by those travelling the waterway itself.

At Hockley Heath, a short arm leads off under an almost ludicrously small but high arched bridge to a weedy, reedy basin and wharf. The old warehouse has been incorporated into the Wharf Tavern. It seems just another abandoned, insignificant small town arm; but for many years this was the terminus, the end of the line. Here the canal lingered while the promoters dashed round trying to find the cash to finish it. So it is not surprising to find that the next section of the waterway seems to have a somewhat different character from what has gone before. The re-started canal promptly plunges downhill through the Lapworth locks. It is an area packed with interest, full of appealing details. The problem of getting a towrope over a bridge, for example, was solved by splitting the bridge. The still comparatively new material for bridge-building, cast iron, was used in a way that would have been impossible with brick or stone. In effect, these are not spans at all, but a pair of cantilevers that do not quite meet in the middle. The locks run down to Kingswood Junction, where once again the canal rested, from 1803 to 1812.

Kingswood has everything to make for a perfectly balanced scene. The narrow locks march down the hill, in their smart uniform of black and white, to a wide basin at the junction itself. There is an essentially practical maintenance yard, with buildings of purplish brick, set off by stone quoins and keystones over the waggon arches. A neat lock cottage completes the group, and there is a touch of nostalgia and wistful romance in the sunken narrow-boats, their working lives over. Beyond this is the southern Stratford, the non-paying, long-neglected section down to Stratford-upon-Avon. Economies show. Cottages were designed on the cheap, with barrel-vaulted roofs. They are very distinctive, now thought of as picturesque, but their main advantage was that they used less material than more conventional designs. The canal turns and twists through an undulating landscape, and where it is forced to leap a gap, it does so economically with cast-iron aqueducts. The first at Wootton Wawen is humble. At the northern end, the hire base has new buildings of

Lechlade, Thames.

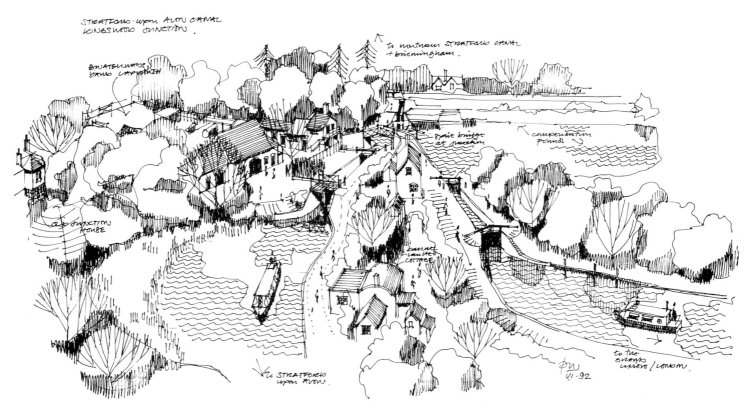

Extending from King's Norton, Birmingham at a junction with the Worcester & Birmingham Canal, the Stratford-upon-Avon Canal travels south through Yardley, Hockley Heath and Lapworth to Kingswood Junction, where it joins its southern section and also the Grand Union (formerly the Warwick & Birmingham). The southern section is famous because of its reprieve from closure, and the splendid efforts of volunteers under David Hutching's direction to restore it to navigation in the early 1960s. Well-known, too, for its barrel-vaulted cottages and interesting split bridges – cantilevered iron – the canal enables boaters to visit Shakespeare's town and link with the Upper Avon Navigation, also restored by David Hutchings.

a very becoming modesty, showing the same unassuming qualities of practicality and sound, unpretentious design that one accepts, almost without thinking, in the older buildings, but finds all too rarely in the new. Wootton aqueduct serves as a prelude to the grander theme of Edstone, designed by William James. It now crosses river, road and railway, the trough carried on a series of narrow brick piers. There are few concessions to gracefulness, and the low-level towpath looks as if it has been stuck on as an afterthought. The aqueduct is long – 475 feet – but at only 28 feet tall at its highest point, it lacks the drama of Pont Cysyllte.

The canal continues to twist through pleasant countryside until it comes to the Wilmcote locks, which drop it down towards the edge of the Stratford suburbs. Then, quite suddenly it seems, the canal bursts out into public gardens by the Memorial Theatre under the metallic gaze of Shakespeare and his creations. Even now, people wander around it convinced that it is an ornamental park and wondering why boats are allowed in. Few canals have a more triumphal final flourish.

This has been rather a long description of the Stratford Canal, but to understand its restoration and what it means, one has to know why it needed rescuing in the first place – and why anyone considered that the effort and expense of restoration were worthwhile. The obvious answer might seem to be that it provides a wonderful cruising route through delightful countryside, but I hope enough description has been given to show that the canal structures themselves add immeasurably to the pleasures. In them, the changing fortunes of the canal can still be read. There is visual pleasure – and also the pleasure of admiring the ingenuity of the canal's builders. If the canal had been no more than a water-filled ditch that allowed boats to move through a pretty landscape, would the work of restoration have been justified? Would it ever have been begun? Perhaps, but it seems unlikely. Other canals would certainly have been left severely alone.

The Ashton Canal cannot have won many friends for the scenic delights of the area through which it passes. There is not much to it: fewer than seven miles of waterway, with 18 locks bashing straight on down through a landscape of mills, factories and warehouses – once grimy with use; now, as often as not, grimy with disuse. This is Industrial Revolution Manchester, Cottonopolis Manchester; once the greatest textile town in the world. The canal was begun in 1792, but in 1793 the proprietors had enough confidence in the scheme to push through another Act, authorizing a number of branches, and more than doubling the length of the original main line. The branches have withered and died, but the main line remains – and remains in use after years of dereliction. So, if the Ashton was not restored for its obvious beauty, what is the explanation? Partly the answer can be found in the Cheshire Ring, a circuit of waterways that does include the very real scenic beauties of the Macclesfield and the Peak Forest Canals, but just as important was the realization that the urban canal has its own set of values. Water has its own attractions. Even a dull building looks twice as imposing when doubled in size by reflections, and ripples can turn its hard edges into swaying curves like a billowing marquee. The visual pleasures of a lock remain the same in a town as they are in the country; and almost no one can resist watching a boat go through. It could be argued that the more depressing the surroundings, the more valuable the canal becomes as a relief. It can be an enclave of peace; the towpath becomes a footpath, where the only traffic is a passing boat. There is another element here as well. The canal represents a particular moment in time; it derives from the period that saw Manchester grow from village to town to city. It was part of a trade route that extended out to the world. Along these waters came cotton from America or India, and coal for the mills. The canal fed the Industrial Revolution that was not merely changing Manchester but was changing the world. So it has a unique importance, which in itself would be quite enough to justify its restoration.

Follow it in either direction, east to its junction with the Huddersfield Narrow or west to the Rochdale, and then continue down either of these canals, and you will find the same message reinforced. The connection between the canals of the mania years and restoration is not always so direct. Dudley, in the heart of the Black Country, was

one of the great pioneering centres of the age. It was here, in 1712, that Thomas Newcomen came to build his first steam-engine – and a full-size replica now puffs away at the Black Country Museum. So it is not surprising to find the region joining the canal age quite early on, with a canal that included a tunnel of monstrous proportions. In 1793, the Dudley Canal Company promoted a new route, from the first Dudley Canal down to the Worcester & Birmingham. It brought new traffic to the Dudley Canal, but at the same time put a great strain on the long, narrow tunnel that connected Dudley No. 2, as it became known, to the Birmingham Canal to the north. Something like 40,000 boats a year were making their way along a route with no towpath and too narrow to allow boats to pass, except at a number of underground junctions and basins. Something clearly had to be done, and there were two alternatives on offer: to improve the old tunnel or build a new one. The proprietors opted for the latter course, and in 1858 the 3,072-yard-long Netherton tunnel was opened. Ironically, the new, improved, short route lost some of its value, when the even longer Lappal tunnel collapsed. So boats going through Netherton then had to wind their way on a meander round the hills to the Dudley No. 1. Most thought it was well worth the trouble. Netherton tunnel represented pure luxury for boat crews, for it had towpaths at either side, and given a choice between that and the old Dudley tunnel running parallel to it, there was simply no contest. The original soon began to deteriorate, and it was closed. It was always a candidate for reopening. Anyone who thinks of tunnels as nothing more than dark holes in a hillside has never visited Dudley.

The early tunnel-builders did not believe in wasting their efforts. If they were going to go to the trouble and expense of delving deep into a hillside, then they reasoned that they would exploit whatever was in there. At Harecastle tunnel on the Trent & Mersey, coal was worked and sent out by boat; at Dudley it was limestone. It was in fact the need to work the limestone mines under Castle Hill that got the whole process started. When Lord Dudley succeeded to the title in 1774, he inherited estates rich in coal, limestone and fireclay, conveniently situated halfway between the Birmingham and the Staffs & Worcester Canals. All that was needed was a connecting link. In 1775 work began on a modest canal, known as Lord Ward's, which has a 226-yard tunnel that ended at underground limestone workings. In time, the canal system around Dudley was extended, with branches heading north and south, but divided by a high ridge of ground. In 1785 approval was given for the little original tunnel to be extended to become the Dudley Canal tunnel, which would end up with an overall length of 3,154 yards: but not immediately. By 1789 work had virtually come to a halt, the engineer Thomas Dadford had left and the cash was spent. A new engineer, the very experienced Josiah Clowes, was brought in, more cash was raised by a second Act and there was a new sense of

Trawler, Caledonian Canal.

Castle Mill Basin at the heart of the complex of tunnels that make up the first Dudley tunnel.

urgency about the proceedings. In 1792 the tunnel was opened, but work went on creating branches to the various mines and quarries. Originally, shafts were sunk at intervals of approximately 220 yards, and three were kept open as air shafts. The tunnel had no towpath, but no one considered the notion that boats might one day go through under their own power. Three air shafts were reduced to one, so that the ventilation is now so poor that no engines, other than electric, can be used in the tunnel. Those who come this way must do as generations of boatmen have done – and leg it. It is not too difficult, as the tunnel is so narrow that the two leggers can lie side by side and 'walk' their way along the tunnel walls.

The northern end of the tunnel is now home to the Black Country Museum. The old basin lies at the heart of the museum complex and is dominated by a massive range of lime kilns, their sheer size giving an indication of the importance of the limestone quarries, while elsewhere on the museum site there are reconstructions of the industries which the canal served. From the site, trip-boats are run into the tunnel. At first it seems conventional enough, low, narrow and dark, with a brick lining pierced by 'weep holes', which allow water to drain out through the rock and into the tunnel, leaving sickly white tongues of calcite lolling out from the holes. Then the tunnel opens out to Short's Mill Basin, and the entrance to the first of the mines: a blocked-in tunnel at the side carried a tramway that brought stone to the underground wharf. Just beyond that is one of the most spectacular features, Castle

Mill Basin. This marks the end of the first tunnel. The boat noses out through a hole at the foot of an immense cliff, and eases into a wide pool, surrounded by rocks and trees. Three tunnels met here, including the now disused Wren's Nest Branch, two-thirds of a mile long, which ended in yet more cavernous basins.

The first part of the journey sets the pattern for the whole. Narrow brick-lined openings, into which the boat slides like a finger into a glove, alternate with sections where the bare rock has been exposed. And every so often there is a great opening out. Sometimes, as at the aptly named Cathedral Arch, a complex of brick vaulting marks an underground junction; at other places old workings appear as limestone caverns. The sheer scale of the tunnel, with its wide and narrow sections, its side tunnels and its mixture of brick lining and bare rock made it a great problem for restorers, who were faced with clearing the approach canals as well as the tunnel itself. A Preservation Society was formed in 1964, when the tunnel was threatened with being blocked off by a new railway embankment, but when that threat receded, the Society turned from the passive defence role to the more active restoration role. In the 1970s massive work-parties were the order of the day, with hundreds of volunteers descending on Dudley for weekend 'dig-ins'. In 1973 the Dudley tunnel was reopened by the

Sounding Bridge at Netherton on Dudley No.2 Canal.

then chairman of British Waterways Board, Sir Frank Price, and the Mayor of Dudley. The story, it seemed, had a happy ending. But the Dudley Canal proved, like the Stratford, that it is not enough just to re-open a canal. It has to be kept open as well.

Many stories have circulated, over the years, of the time when the tunnel was open to anyone prepared to take a boat through without using the engine. The tunnel really is very narrow in parts, and a gauge at the entrance shows the maximum size of boat that can expect to get through. One jolly band of holiday-makers ignored this and charged merrily on, until they found themselves wedged and immobile. They put messages in bottles and threw the bottles overboard. They waited for rescue: no rescue came. They were soon surrounded by bottles bobbing about all around them, but going nowhere. At last one brave soul had to volunteer to swim out: one hopes, having seen the water in this area, that he kept his mouth shut. Sadly, it all had to be closed again, but on this seesaw of good and bad tidings, the balance has tipped again and once more Dudley is open to the intrepid and energetic. There are restrictions, and all boat movements are controlled by the tunnel-keeper. But at least there is no fear of being stranded, as every boat is counted in and checked out at the far end.

Tunnels represent some of the greatest obstacles facing restorers. Sapperton (see p.40) is an obvious example, and the restorers of the Huddersfield Narrow Canal have the biggest tunnel of them all. Standedge tunnel, at 5,415 yards, dwarfs even the Dudley, and the length of time it took to construct makes the Dudley's seven years seem almost immodestly quick. Begun in 1794, it was very much a creation of the mania. It owed its existence to two other waterways. In 1774 Sir John Ramsden had built a broad canal, in effect a branch of the Calder & Hebble that linked Huddersfield into the north-eastern network of rivers and canals. Then in 1792 work began on the Ashton, at which point it seemed a good idea to link the two together. Geography, however, was not on the side of the canal-builders, for between the two lay the Pennine hills. The engineering works were to be on a vast scale. Numbers are impressive, even more so when put into context. There were 74 locks in all, a daunting number for any canal, but especially for one just under 20 miles long, of which over three miles comprise tunnel. It averages out at a lock every 400 yards. The tunnel would have come as quite a relief to the boat crews, if only it had had a towpath. As it was, it had to be legged, and as time went by things got worse rather than better. In the 1840s the railway arrived and a new tunnel was built alongside the canal. It kept the canal company busy, for short cross-tunnels were built so that the spoils could be carried away by boat. This was fine – until the railway opened. Then, whenever a train went through, clouds of smoke and steam billowed out into the canal tunnel. Anyone travelling the canal today would have the even less appetizing aroma of diesel fumes to accompany the journey. When the canal was still in the planning *Elland lock, Calder & Hebble.*

stage, there were proposals for avoiding the tunnel altogether by joining the two halves with a tramway. In retrospect, that seems quite a sensible notion.

The Huddersfield Canal is a very interesting case, in terms of restoration. When work first began on bringing old canals back to life, the case for their revival seemed reasonably clear-cut. The Stratford appeared to have everything the holiday-boater could possibly want. But what about the Huddersfield Narrow? Put in bald terms, it fights clear of the urban environment of Huddersfield itself, but never shakes off its industrial origins. The mills continue to punctuate its progress, their chimneys a series of exclamation marks signifying the end of each phase. There is no respite from the efforts of locking, other than a dark, dripping passage under the Pennine hills. The charms of such a canal are not immediately apparent, but they are there, waiting to be discovered. Perhaps 'charm' is not the right word, suggesting old thatched cottages, roses round the door; fields of corn splashed with poppies; a gentle swell of grassy hills. The countryside through which the Huddersfield passes has none of this gentle softness. The valleys close in, rearing up to either side to dark margins; black lines of crags define the horizon. The dark gritstone is ideal for use in big, rough blocks: it does not lend itself to delicate, decorous carvings. So something of the toughness of the countryside is reflected in the buildings. It is an area with a character very much its own, still tied visually to the wool trade on which, in the canal age, the region thrived. Sheep graze the hills; many of the older houses still have the long upper-storey windows that mark where weavers once worked at their hand looms. But down in the valley bottom the steam mills predominate, with their conspicuous tall chimneys. Coal for these mills was the staple trade of the canals: this was the trade that brought them into being. In the Colne valley, on the Yorkshire side, the canal slides between its own reservoirs and the almost equally imposing mill pond, emphasizing the close relationship between transport and industry. A row of workers' cottages completes the scene, making for its own mini-settlement drawn away from the road, higher up the slope of the valley, by the enticement of cheap water transport. Sometimes the canals seem so far removed from the world they were built to serve that it is difficult to believe they were ever anything other than routes for pleasure-boats. This is certainly not so in the case of the Huddersfield. Here the context is clear: hill farm, mill, moor and canal make a most satisfying whole.

How many people will want to use a canal between Huddersfield and Manchester, which involves working 74 locks and the longest tunnel in the country? There will always be the enthusiasts who want to travel every canal, and who measure the success of a day's travel in miles covered and locks worked. There will be others who respond to the special environment of this waterway and who appreciate its history. But are there enough of these to justify the effort and expense of restoration? If restoration could be justified only in terms of boating, then the answer would probably be 'no'. But that is not the only reason for restoration. As one who has known the canal as a wholly derelict waterway and visited it at least once a year since restoration work began, I have become aware of the tremendous positive impact it has had on the countryside through which it passes. The sullen, grimy ditch has become a real waterway of ruffled water; the ugly, concrete-capped lumps of locks have emerged as beautifully crafted stone chambers, while the regular march of black-and-white balance beams down the valleys have brought a crispness to the perspective. To walk the towpath is vastly more pleasurable now than it was before. Soon, one hopes, the regular passage of boats will enliven the scene and play an important part in keeping the channel clear and weed-free. But even when the canal is back in use, for every one who comes this way by boat, there will be many more who will enjoy the canal's role in defining the scenery and history of the region.

The same arguments can be applied to the other trans-Pennine route currently undergoing restoration. The Rochdale Canal was another product of the mania years. Delayed by two unsuccessful attempts to get Parliamentary approval in 1792 and 1793, with John Rennie as its guiding hand, it finally won its way through in 1794 when the indefatigable William Jessop took control. The sticking point was water supply. The local mill owners were understandably concerned about any scheme that threatened their own essential supplies, and Jessop was able to overcome their fears by proposing to provide all the water the canal needed through reservoirs. The Rochdale is a majestic canal, charging across the Pennines via broad locks, 74 feet long and 14 feet 2 inches wide. The difficulties faced by the engineers were immense. Yet this canal, begun in the same year as the Huddersfield, was opened in 1804, well ahead of the Huddersfield and beating even the Leeds & Liverpool, begun 24 years earlier.

The Rochdale is both like and unlike the Huddersfield. It is similar in that, for much of its way, it passes through the Pennine hills, and offers the same alternation of open moorland and tight-clenched mill towns. It too uses local materials and local styles for its buildings. And it too found itself with an unwelcome neighbour, the Lancashire & Yorkshire Railway, which was to prove an enthusiastic competitor. Yet the visual effect of the two canals is very different. Where the Huddersfield takes a direct line, the Rochdale starts off heading west towards a gap in the hills, and then swings right round to the south en route for Manchester. This did not decrease the number of locks required. In fact, the Rochdale outdoes the Huddersfield by quite a margin – 93 locks in all, squashed into 33 miles. It did, however, mean that Jessop could dispense with a long summit tunnel – and given the troubles that he faced on the Grand Junction, this must have been a great relief to him.

This family frozen in on the Rochdale Canal would no doubt have been astonished to hear that, nearly a century later, the canal was being restored for pleasure-boating.

The appeal of the canal is one of contrasts: large- and small-scale; town and country. At the eastern end, at Sowerby Bridge, it joins the Calder & Hebble, which explains the size of the locks. The boats on the latter are wide-beamed barges, which the locks were built to accommodate – but they had a maximum length of 57 feet 6 inches. The barges could use the canal; but narrow-boats coming from Manchester had to stop at Sowerby Bridge. So, as happens whenever two incompatible systems meet, a large basin with warehouses was developed. The warehouses, built over arches for easy loading and unloading, create an effect rather like a smaller-scale Ellesmere Port. There is still a sense of grandeur and importance on display. And the basin is still in use, home to a fleet of hire boats. Warehouses and mills pop up along the route, and they all have this same air of un-selfconscious dignity. When something is done on the large scale, it never fails to impress. I am not sure, however, that the small-scale delights are not, in the long term, more satisfying. The local stone that began as solid, rugged chunks wears and weathers so that in time its edges are smoothed, its surface shaped and curved. You can see the great variety of ways in which stone can be used at the bridges. Bordering the towpath there is likely to be a dry-stone wall, that miracle of balance, which seems to require the skills of a jigsaw-puzzler rather than those of a builder. The bridge itself will be built of rough, squared blocks, but arch and parapet will be emphasized by more meticulously dressed stone. Often there are steps up to the bridge, so worn by feet that they seem to sag under the weight of the years, their treads dipping in gentle curves. The tiny hill streams that trickle down from the moorland are not allowed simply to run to waste, but are added to the canal. But first they run an obstacle course of gulleys and troughs, so that the sediment settles and is not carried over to block the waterway.

The route alternates between open countryside and towns. At times it takes an easy route (easy, that is, by the standards of a Pennine route); but where it must force its way through the spine of the hills, the effort shows. Near Hebden Bridge the river has cut its way through a deep valley, above which the hard edges of gritstone crags leave no illusions about what lies just below the surface. Through this valley the engineer laid his route and the men blasted and hacked their way. Stones were levered into place to prop up the river bank, and a narrow ledge was cut for the towpath, leaving enough space for the canal but precious little more. Once past Littleborough, however, everything changes. Stone and Pennines are left behind, the views widen and the main features are now the livid red-brick mills with their tall chimneys. It is a prelude to the gradual closing in of the urban world, as the canal heads on towards Manchester.

The older mill towns – the stone towns – have, somewhat surprisingly perhaps, become rather fashionable. Hebden Bridge is the most obvious example, and the canal gives a view of the whole story of its development. Up on the hill is the old settlement of Heptonstall, where the hand-loom workers earned a living. Then came the mill age, and what was needed first was water to turn the wheels, so the emphasis moved down to the river in the valley below. Hardly had the water-powered mill been established when the steam engine came along to offer a more efficient alternative. Now the canal came into its own, as the coal barges and narrow-boats plied their trade. There was a great outburst of mill building, which left all too little space for homes and people in the steep-sided valley. No room here even for the familiar back-to-backs of other cotton towns. Instead, Hebden Bridge became famous for what one might call top-to-bottoms. One terrace of houses was built on top of the other, the upper row having entrances on one street and presenting a conventional two-storey face to the world; the lower with its entrances on the next street down and showing all four storeys. So, from the canal, Hebden appears as towering terraces, rising above the mill chimneys. It makes for an unusual and spectacular urban landscape. And the canal, restored in this section, greatly enhances the appeal of the area.

One of the most remarkable features of the restoration scene of the 1990s is its diversity. There are canals spread right across the country that have been restored, on which work is well advanced or is still just getting under way. Nothing, it seems, will deflect the devoted restorer. Canals which a few years ago would have been considered beyond all hope of salvation have been taken on. The restoration of the Stratford Canal was seen as a major and very challenging undertaking not so many years ago, yet it was a canal that had never officially been abandoned at all. Since then we have seen such major waterways as the Basingstoke and the Kennet & Avon brought back to life. The Kennet & Avon, however, also tells another story. Restoration work began on the first lock, at Sulhamstead, in 1965 and in 1990 the Queen came to the immense flight of 29 broad locks at Caen hill outside Devizes to declare the whole canal officially open. Then in 1992, before the holiday season had even got under way, the through route was closed again; and padlocks were placed on the lock gates. The winter of 1991–2 had been unusually dry, and not enough rain fell in the early months of 1992 to fill the reservoirs. There was nothing to be done: Caen hill locks, so triumphantly reopened, were now rather quietly closed again. There are ways of overcoming the problem, but they cost more money – a great deal more money. The volunteers whose efforts had, they thought, led to a resounding victory found that they had to get back to work again. The probable solution is a pump-back system. As water is pushed down through the locks, so the pumps will shove it back up the hill again. Just a million or two pounds should see the task complete. It is a classic case, demonstrating that no one can ever sit back and say that the work and expense are over at the grand reopening. We now know that all it represents is just the end of the first phase.

There are canals where the first phase offers quite enough in the way of problems, without anyone concerning themselves with the second stage. The Wilts & Berks has disappeared for a great part of its length. One section has wholly vanished under the mushrooming town of Swindon, the connection with the Thames has long since gone, the final section at Abingdon having disappeared beneath a housing estate. Some parts no longer register as having a canal connection at all. The A420, Oxford–Swindon road divides near Shrivenham to pass beneath the twin arches of a railway bridge. What is not at all obvious is that when the railway was built, it did not pass over two halves of a road: one arch covered the highway, the other the canal. In spite of all these difficulties, the canal society is pushing on with the long process of restoration. Is it a feasible plan? The first reaction would be to say 'No', but so many schemes that once would have seemed to be impossible dreams have now become practical realities that it is a brave soul who now declares anything to be impractical.

Not every canal can be restored; many would argue that not every canal should be restored. Yet even those that may never see boats again do not simply vanish from the landscape. The Somerset Coal Canal is a case in point. It is another of that group of canals built to serve a very specific function. It was not part of any through route, but simply existed, as its name suggests, to feed coal from the once-thriving Somerset coalfield into the broader canal system. Once a railway system was established that could do the job as well, if not better, then there was simply no trade available that could keep the canal alive. But that does not destroy the historical interest of the old route. This was a canal built through seemingly impossible terrain of undulating hills and steep-sided valleys. The usual array of locks, aqueducts and tunnels was brought into use, but more was needed. The first response was the creation of a vertical lift (see p.172): the next was to build an inclined plane; and a third solution was to build a flight of locks that went through a hairpin bend to climb the hillside at Combe Hay. The details of the canal are as fascinating as its changing patterns: near Combe Hay, for example, one can find the gently decaying remains of iron lock gates, while close to Caisson House is the weedy, overgrown basin that once stood at the top of the incline. At the time of writing there are no plans to put boats back on to the canal, other than on the short section near the Dundas aqueduct, now in use for pleasure-boat moorings. A good deal of the towpath, however, survives as a footpath. In 1992 a Somerset Coal Canal Society was formed with the aim of preserving the main features of the canal. The towpath walk is a delight, passing through a beautiful rural landscape. Anything that can be done to preserve and, just as importantly, to explain the canal features can only add to the interest of the area.

Canals have gone through a whole range of development. In their infancy they were the wonders of the age, the transport routes that made the Industrial Revolution possible. In time they were overtaken by an even more efficient transport system, as the railway network began to spread. Gradually, cargo-carrying gave way to a trade in holiday-makers, and as the leisure opportunities offered by canals began to be appreciated, so the impulse to restore grew ever stronger. It is a long history, stretching back now for more than two centuries. The reasons that caused investors and speculators to scramble for canal shares in the 1790s are no longer valid in the 1990s, but the canals survive. The questions now are: have the waterways of the eighteenth century a future in the twenty-first century, and if so, what is that future to be?

OVERLEAF: *Stourport, Staffs & Worcester.*

LOOKING AHEAD

There is a tendency among some people to write off canals as quaint anachronisms, of interest to no one except a few eccentrics in bobble-hats, who like messing about in boats. To counteract that view one can turn to the statistics. Of course, everyone knows the old story about lies, damned lies and statistics, but there are times when the figures can give generalizations a real significance. So here they are.

Canals are only of minority interest? Well, 50 per cent of the population lives within five miles of a canal; there are 8 million casual canal visitors every year; 150,000 people hire boats for holidays; and 120,000 anglers pursue their solitary pleasures there. As for the assertion that canals are at best no more than mucky ditches, best used as a repository for unwanted pushchairs and plastic fertilizer bags? There are 62 Sites of Special Scientific Interest (SSSIs), 2,050 listed structures – structures deemed worthy of official preservation – and 135 ancient monuments situated alongside canals. But what, say the cynics, about the hard cash? Well, if nothing else, having a canal alongside your property is now seen as increasing its value. But perhaps the most surprising statistic shows that freight traffic has not died – it had reached almost five million tons a year at the last count.

What these figures suggest is that there are many different ways of looking at Britain's waterways. They can be seen as a valuable part of a growing leisure industry, as important wildlife habitats or as an extra line of passionate prose in an estate agent's or developer's brochure. But the obvious starting point has to be the reason why the canals were built in the first place. They were designed to carry cargo. The question, is will they do so in the future? And, if so, will trade increase or decrease?

It may seem slightly perverse to begin a look at the future with a return to the past, but it is decisions taken two centuries ago that shaped the system we have today and which, to a great extent, limit the options available for development. Some of the 1793 canals, such as the Stratford, have no realistic future as freight carriers. Others

The harbourmaster's 'pavilion' at Limehouse, designed by Peter White and John Holland, surrounded by new houses.

may have a small role to play. It was a delight to see a motor boat and butty, both loaded with coal, coming down the Grand Union in 1992, but it was a delight born of nostalgia for something that has, in reality, gone. There will be no great fleets of working boats on that canal to match those of the early years of the century.

The last of the 1793 routes, the Stainforth & Keadby, is a very different case. It was to become a link in a successful trading system, joining the River Don to the Trent and, through the Trent, to the ports of the Humber. The scene at the entrance to the canal at Keadby is still suggestive of busy, commercial traffic. Massive lock gates separate the still waters of the canal from the turbulent flow of the tidal Trent. Near the lock entrance itself, a new jetty on the river is busy with small coasters loading and unloading. This is all to the good, but unfortunately the jetty seems to have disrupted the river's natural flow, as it rises and falls with the tide. As a result, the build-up of silt has turned the canal the colour of canteen coffee. This only emphasizes the pattern of use: the river with its enormous lift bridge carries a modest but steady stream of coasters and large, motorized barges; the canal's traffic has dwindled. But there are ample signs of its importance in the past – indeed, the recent past. Even railway traffic was subservient to the canals, with an odd bridge built on the skew over the canal, which neither lifts nor swings, but slides smoothly away to allow boats to pass. Road bridges are more conventional moveable bridges.

The canal runs straight over a flat landscape, with little in the way of interruption. Crowle Wharf is a sad area, permeated with a sense of decay. Even when new it must

Humber keels under sail on the Stainforth & Keadby Canal at Crowle. The lee board can be seen drawn up on the port side of the hull.

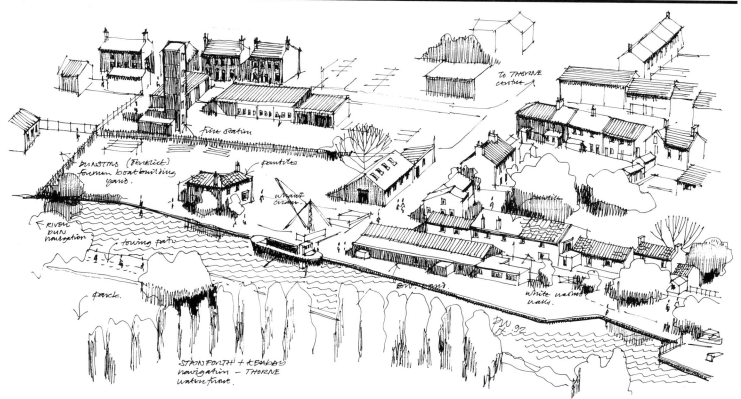

have been more than a touch drab, with stabling and warehousing in a dull red brick. The railway alongside still functions, but is scarcely more cheerful. Thorne brings the waterway travellers who come this way their one moment of excitement – the lock, the sight of a still-active maintenance yard and a good deal less active wharf area – and the latest addition to the scene, a marina. Originally the canal ended at Stainforth, where it joined the River Don, and where the local colliery kept the boats busy. Though this marks the end of the 1793 canal, there is very little to show that this is the case, apart from the old lock, half-buried under a high flood bank. The main line simply seems to go on, until Bramwith lock is reached. With its double gates – so that it can be used as a short lock to accommodate small craft, or full length for the bigger barges – the lock is undoubtedly impressive, though its old hand-operated paddle gear and winches for opening the gates are cumbersome. Overshadowing the whole scene are the massive, yet delicately curved, cooling towers of Thorpe Marsh Power Station. The conjunction of power station, colliery and canal was for a long time the mainstay of traffic in the north-east of England. Beyond the lock, the New Junction Canal appears, a latecomer, begun only in 1896 but offering a direct connection to the Aire & Calder. The route to the south now continues via the Don to the Sheffield Canal, begun in 1815 and offering a through route from the heart of that city to the Trent. That system eventually became amalgamated as the Sheffield & South Yorkshire Navigation.

This waterway joins the River Don Navigation at Stainforth, and proceeds via Thorne, Medge Hall and Crowle Wharf to Keadby on the tidal River Trent. Some 12 ½ miles long, it still carries freight traffic. Just below the lock at Thorne there was once a thriving waterfront, full of boat-building and associated activities. Rope Walk, the pubs and several quayside warehouses remain, but the rest is sadly deserted and derelict; together with the closure of Thorne colliery, it conveys an impression of despair. But with open space opposite and an enterprising scheme to link the town back to its waterway, Thorne could rediscover its former prosperity.

Back in the 1970s, the Sheffield & South Yorkshire seemed ripe for development. The need for improvement was obvious to all who came this way by boat. I recall waiting at Sprotborough lock while a tug brought three barges down from Rotherham. First the tug and one barge squeezed in and were locked through in a conventional manner – with everything hand-operated, one might add. Then the lock was refilled, and the top gates and the bottom paddles were opened to suck barge number two into the lock to start the journey. The same process then had to be repeated for barge number three. Finally, when all the barges had been reunited, the journey could continue. It is not hard to imagine what a transport manager, looking for an efficient, modern way to move his goods, would have thought of such a scene. Improvement was costed at £3.5 million: it sounds like a great deal of money, but this was at a time when expenditure on the roads was running not into millions but, literally, into hundreds of millions. The government dithered, the costs rose, and in an all too familiar scenario, improvements came too late to catch the market. Scarcely had the tablecloths been cleared, the ashtrays emptied and the glasses washed after the official celebrations than things began to turn sour.

One argument in favour of the improvement of the waterway had been the need to service the mighty steel industry of Rotherham. Then came the new philosophy of the 1980s, which declared that market forces would decide who should prosper, who should fall. And if those forces decreed that heavy industry no longer had a place in the British economy, then so be it – the industrial world would have to go, and the phoenix that was to rise from its ashes would be the new Britain of service industries. Travellers on the M1 have, over the last few years, been able to enjoy (if that is the word) a bird's-eye view of the system at work. They have seen the steel works destroyed and in their place the massive Meadowhall shopping centre has grown up. The waterway is still there, but its potential seriously diminished: supermarkets and clothing stores seldom show much interest in barge transport. It is difficult to say if this sequence of events is an argument for or against modernization of the broad waterways of the region. Some would say that heavy industry is in decline and likely to decline further, so money should not be wasted. Others would argue that the country needs a sound industrial base, and that will rely, as it always has done, on appropriate transport. So if we assume that we are not going to live up to Napoleon's scornful phrase and actually become a nation of shopkeepers – it sounds trendier in French, *une nation de boutiquiers* – and that there is to be an industrial future, what part should the waterways play?

The argument in favour of waterways has always rested on their energy efficiency. Back at the beginning of the canal age, engineers worked out the load that could be moved by one horse by different means and came up with one-eighth of a ton for a pack animal, two tons for a waggon on a good road, eight tons for a waggon on rails and 50 tons for a boat on a canal. Substitute one-horse power for one horse and you have an encouraging picture. Over the years, road transport may have got better and more efficient, but it is still true to say that you would expect to use four times as much fuel in moving goods by road as you would by water. If the economics are that simple, then one would expect to find a rush of goods on to the canals. But there are arguments on both sides. On behalf of the waterways, there is the very powerful argument that every ton carried by boat is a ton not clogging up our already overcrowded roads. Barges seldom run into people, cause traffic jams or shake old buildings to pieces with their vibrations. Road transport, however, has always had one immense advantage over all other alternatives: it can offer a door-to-door service. A truck can also, in spite of traffic congestion, travel a good deal faster than a boat, so that any extra fuel costs are more than offset by the saving in wages. These are undeniable powerful arguments, which in many, if not most, situations will win the day. There are, however, a not insignificant number of instances when water transport should be able to compete. One way to overcome the cost problem is to go for size. One man controlling a barge carrying 200 tons that takes five hours for a journey costs the same as five men taking one hour each, but each carrying only 40 tons. The new generation of motorized barges, or systems using push-tow, provides just those economies of scale. It is not a new notion.

A train of Tom Puddings snaking down the Aire & Calder at Knottingley.

The Aire & Calder was the scene of a series of increasingly sophisticated experiments using steam power. Tugs pulled trains of barges from as early as 1831 and their obvious success created a demand for ever larger vessels, which in turn called for bigger locks. The new port of Goole had been established in the 1820s, and there was every prospect of a vast coal-exporting trade from the collieries of South Yorkshire. In 1860 the first of a new generation of locks was built, at 206 feet almost three times as long as their predecessors. It was like a game of technological leapfrog: an improvement in boat technology called for an improvement in the waterway itself which, in turn, made new advances in boat design possible. It was William H. Bartholomew, the Aire & Calder's company engineer, who brought together the different elements, to create something quite new for the waterways. Right up to the end of trading days on the narrow canals, coal-carrying was still largely a muscle and shovel job. Bartholomew realized that, given a new port with new facilities and a greatly improved waterways system, the vessels could be tailor-made for the job in hand. The result was the Tom Pudding train. The name comes from the appearance of the 'pans', the small barges that did look very much like overgrown pudding tins. They were oblong boxes, holding roughly 35 tons each. In the original version, a string of these pans was made up and a false bow added at the front. Cables ran the length of the train, from the bows to the steam-tug sited at the back. By pulling on the lines, the straggle of boxes could be bent to go round corners. It was ingenious, but not very successful, and a more conventional system was soon adopted. The Tom Puddings represented an early example of push-tow on the waterways. It was also an example of bringing together the technologies of the dock system and the vessels using it, much as a modern container port does. The pans were individually filled at a waterside colliery, then taken to Goole, where special hydraulic lifts picked up the pans and up-ended them into a waiting coaster. A similar system was to be used on a larger scale in the twentieth century for unloading coal barges at the Ferrybridge Power Station.

Push-tow is a system beautifully suited to wide waterways, and can be seen in its most dramatic form in America on the Mississippi. Here as many as 30 barges, with a total capacity of as much as 59,000 tons, can be linked to a single push unit. The 'tug' in fact has something like the proportions of a small block of flats, from the top of which the steersman peers out towards the distant front of his great raft of barges. The Aire & Calder is scarcely the Mississippi: a photograph taken from the bridge at Natchez of a monstrous push-tow would make it look like a canoe on that wide American river. It does, however, help to make the point that today all kinds of combinations, of almost any size, are possible. But even the biggest push-tows are limited to inland waterways. That does not mean, however, that barges cannot go to

Dunardy, Crinan Canal.

sea: it just requires a little more ingenuity to get them there. Once again, this is not a new idea, even if modern engineers have come up with answers that would have astonished – and no doubt delighted – their forebears. To put modern achievements into context, however, one has once again to step back in time.

There is a long tradition in Britain of river craft that were also capable of short coastal runs. Perhaps the best-known examples are the spritsail barges of south-east England, popularly known as Thames barges, though they were widely used all up the East Anglian coast. Quite a large number of these magnificent vessels with their traditional red sails survive, and they meet regularly to preserve an old tradition of barge matches. On race days, there will be a crew of four or five to work the boats, but in their working lives they were worked by just two – and towards the end of trading, when cash was scarce, it was by no means unknown for a sailing barge to be worked single-handed. They are superbly rigged to keep as many operations as possible close to the hand of the helmsman. Like other river craft, they are flat-bottomed, but overcame the lack of a keel when going to sea by the use of lee boards. These great, pear-shaped slabs of wood are suspended on either side of the vessel and are alternately raised or lowered – depending on which tack the barge is on – to prevent the vessel from skittering sideways over the water.

The river systems of Britain bred their own distinctive type of sailing barge – Severn trows, Mersey flats, Norfolk wherries, and so on. In the north-east it was the Humber keel or its close relation, the Humber sloop. They have a similar hull, but the keel is square-rigged, while the sloop has a fore-and-aft sail pattern. In the keel you can see a vessel which would have been perfectly well understood by the Vikings who raided this coast centuries before, and which any competent medieval sailor would have been happy to handle. Like the Thames barge, it has lee boards for going to sea – and a trip to the mouth of the Humber in bad weather can seem like a visit to the North Atlantic. But these vessels also traded far inland. They were regular visitors to the Sheffield & South Yorkshire, sailing through the fields and past the collieries when there was a favourable wind, and being pulled by horses from the towpath when the sail had to be lowered. The men in charge were known as 'horse marines'. In later years, tugs would be used for haulage.

The sailing barges were wonderfully versatile, but they had their limitations. Today Britain is politically and economically tied to continental Europe, as she has not been for many centuries. Look across to Europe and you see a far more flourishing waterways trade. Locks have been enlarged to take a new generation of giant push-tows; huge lifts have been built to replace locks altogether – totally dwarfing Britain's 'wonder of the waterways' at Anderton. British Waterways introduced push-tows in the 1970s, and there was a real glimmer of hope for a scheme that would see barges running (without any need for transshipment of cargo) from the heart of England to the centre of Europe. There were a number of designs, all based on the same idea of barges using the inland waterways, then being loaded, cargo still intact, into a mother ship which took over for the sea crossing, to be released on the other side. They came with exotic names – LASH (lighter aboard ship) was followed by SPLASH, a self-propelled lash. One of the more interesting versions was the Danish-built BACAT, specifically designed for runs from Rotterdam to the Humber, where its barges could be released to travel to Leeds along the Aire & Calder, to Rotherham on the Sheffield & South Yorkshire, and to Gainsborough on the Trent. In this version, the mother ship was twin-hulled, with some barges carried on deck and others locked in place between the hulls, hence the name BACAT 'barge aboard catamaran'. There was nothing wrong with the design, but it failed because of opposition from vested interests, in particular the Hull dockers, who could not face the notion of cargo by-passing the port. In the event, BACAT was withdrawn – but this did not save the jobs of the Hull dockers.

Unloading coal barges at the Manchester Corporation power station at Barton.

The stalwart efforts to increase the carrying trade in recent years have had mixed results. There have always been those who have argued in favour of water transport, if only because the ruinous effects of an over-reliance on road transport are all too visible on every hand. The market is still there to be exploited, but it is very difficult to predict quite what that market will be in the future. Traditionally, waterways have played an important role in supplying coal-fired power stations, but now that industry, too, has been called into question. It is simply no longer possible to rely on well-established cargoes such as coal. So there has to be much greater flexibility, much greater ingenuity in finding new trade. The key word now is 'opportunism'. This means looking hard for cargoes that are available in the right place, and which can be carried along suitable routes. An excellent recent example comes from the north-west of England, where the local chemical industry is producing chlorine for Ireland. This is now loaded directly on to coasters at Runcorn and then sent on its way across the Irish Sea. British Waterways is also busily looking for ways to revive trade with Europe, to link into the extensive continental waterways system. The trade of the future will be different from the trade of the past, but water transport should always have a part to play in any rational transport system. Transport need not, however, mean simply loading cargo on to boats. The canals can provide a route for information, power, and even for water itself.

The 1990s brought a recurrence of an old problem – water shortages. It is not a case of Britain being a particularly arid region (far from it) but, unfortunately, of rain refusing to fall regularly over the whole country. It may be tipping down in the north while there are near drought conditions in the south; there may be a hosepipe ban in Kent while umbrella sales are booming in Lancashire. The idea of using canals to facilitate water supply is not new: Bristol Water, for example, has been using the Gloucester & Sharpness for years. Now, however, a much larger, far more imaginative scheme is being considered. Because the canal system links so many different regions, it would technically be feasible to use it to transfer water around the country. With luck, there might even be enough left over to fill the recently drought-stricken Kennet & Avon! And the concept of the canal system as a network to be used in more than one way has given birth to another novel scheme: it can be used for fibre-optic communication systems. The cables can be laid along the towpaths, with little interruption to canal traffic and none at all to the roads – no holes, no cones. More importantly, perhaps, the administration costs for the optics company can be slashed. Instead of having to deal with a multiplicity of landowners, local farmers, public bodies and a whole range of other interests, the entire deal could be concluded by one set of negotiations with just one body, British Waterways. And what applies to fibre-optics can equally well be applied to gas pipes and electricity cables. The canals that

Wolverley, Staffs & Worcester.

carried coal – the fuel of the Industrial Revolution – can thus be said to have been adapted for carrying the new fuels of a later age.

It would be too much to say that there is an obvious glowing future for commercial waterways traffic. Freight movement, at just under five million tons a year, represents only a fraction of the country's trade. It has to be put into context: there are far fewer miles of freight-carrying waterways than there are miles of road. But put another way, waterways carry five times more freight per mile than the railways, and a surprising nine times more freight than the roads. It seems that there is great scope for improvement – many of us are still convinced that this is true – but somehow the reality always seems to fall short of the expectations. In the past, revival plans, such as the refurbishment of Sharpness docks and the Sheffield & South Yorkshire, have not led to the glories of a new dawn. There was a time when the Green movement seemed likely to have a real effect on government policies, but politicians are almost universally inclined to agree loudly and enthusiastically that action is needed to save the environment, but to show no inclination whatsoever to do anything about it. But the problems of overcrowding, diminishing resources, ozone damage and the rest will not be solved by talk alone, and will not go away.

Economic fashions may change but the facts remain the same. Boats are efficient, cause no harm to the fabric of towns and cities, and inconvenience no one. People groan at the sight of yet another truck; they smile when a boat comes round the bend.

There is one boom area in the use of waterways – namely, leisure. First thoughts turn inevitably to boating, which is, after all, the activity for which the canals were built in the first place, even if the original promoters had in mind cargoes of pig-iron, coal and grain, rather than Mum, Dad and the kids. But boaters are by no means the only people who go to the canal to relax. Anglers throng the banks in their thousands – on match days, the towpath looks like a sale display of green umbrellas. And more and more walkers are beginning to see the towpaths as valuable long-distance routes: just try and imagine any other footpath that will do what the Grand Union towpath does – take you from the heart of London to the centre of Birmingham.

But it remains true that, in terms of business and revenue, holiday boating occupies the main role. It is also true that it is boating that keeps the canals alive. Take away the boats and soon the reeds and weeds spread, the supermarket trolleys and bags of rubbish accumulate, and the canal becomes a place of little interest to anyone, bar a few dedicated hunters-out of moribund waterways. So any view of leisure use must start with boats.

Tens of thousands of people must have done what I did many years ago – taken a casual decision to go on a canal holiday and become an instant addict. Canal holidays are sold as get-away-from-it-all breaks with a difference, which they most certainly are. They are also sometimes sold as carefree, trouble-free holidays which, in my view, is somewhat more debatable. I am not suggesting that the engine is going to break up or the boat spring a leak, but the best laid schemes of mice, men and holiday-boaters . . . . A late spring holiday once took a distinct turn for the worse when I woke up to see a few inches of snow and a near-blizzard blowing. In those circumstances there is nothing much one can do other than make the best of things. The same applies to other, less dramatic, trials and tribulations.

Indeed, I would go further and say that one of the aspects that makes a canal holiday special is the sense of achievement that comes with it. Anyone taking a boat up, let us say, Hatton locks knows that by the end they have done a good job of work. I have always thought of that as being part of the fun, and that those who want absolutely everything done for them might be happier staying at home, steering a computerized boat down a video canal. At the other extreme are the enthusiasts who deliberately seek out difficulties as a challenge to be overcome, ending a holiday on the far reaches of the Birmingham Canal system with the sense of triumph usually reserved for those who have recently climbed Kanchenjunga. Most of us lie somewhere in between. We enjoy the work, the pint well-earned, the contrast between effortlessly chugging along a pound and the burst of activity involved in throwing lines, working gear and pushing gates. Children almost invariably find that it is the steady interruption of locks that keeps them interested, and which gives them the chance not just to be active but to make a real contribution to the holiday.

Canal travel can seem a complicated affair – especially for beginners. We all started off boating knowing nothing, desperately muttering to ourselves, 'If I want to go left, I'll push this thing to the right.' And precisely because working a boat well, and managing a lock efficiently, are skills that can always be improved, we find the pleasure of getting them right all the greater. Even after years of boating, things can still go wrong. I remember on one occasion asking our youngest to hold on to a line, and in a fit of over-enthusiasm he tied it round his middle to make sure he was holding it tight enough. As a result, when I began to move the boat forward, he was dragged along the towpath and was almost turned into a reluctant water-skier.

There will always be a few people who will find the problems not worth solving, the effort too much trouble – but they are a minority. Most of us actually find it is the problems that stick in the mind – as good anecdotes to tell later over a pint. If people ask me which is the most memorable journey through a flight of locks, I do not think of the day we went up Tardebigge and everything worked perfectly – I remember Wigan, when the rain did not so much come down as come across, a miserable, horizontal hosepipe of a storm, which numbed one to the bone. And I remember how warm the fire was in the bar afterwards and how good the beer tasted.

ABOVE: *Mile post, Lancaster Canal*
OPPOSITE: *Leamington Spa, Grand Union.*

There is a difference between the unpredictable natural elements and avoidable nuisances – the paddle gear that is not well maintained, the gates that stick, the swing bridges that will not swing. These are simply irritations. The good news for boaters is that British Waterways is aware of the problem and has made plans to solve it. The first essential is to tackle the urgent maintenance work that has built up over the years, and in 1992 a programme was put in hand to cover all this work within four years. Another, altogether admirable aim is to ensure water supplies, so that nine years out of ten could be free from restrictions. Add to that a determination to improve towpaths, so that waterways can be enjoyed as much by walkers as by boaters, and it seems a wholly praiseworthy plan.

None of these objectives would have seemed controversial over the last few years and decades, and yet they have not been achieved. The problem has not necessarily been lack of goodwill but cash – not to mention a host of other problems, of which most of us are wholly unaware. New anti-pollution laws, for example, mean that the muck dredged from the bottom of urban canals is no longer just muck but, officially, 'toxic waste'. This means that the slurry can no longer just be dumped on the bank, but has to be carried off to special designated sites. This is an altogether laudable arrangement, but it costs money.

And one has always assumed that when a towpath was allowed to go to rack and ruin, this was simply a question of British Waterways falling down on the job. In fact, when British Waterways was formed, although it was given the specific job of looking

Plans drawn up by Waterway Environment Services for the development of Limehouse Basin.

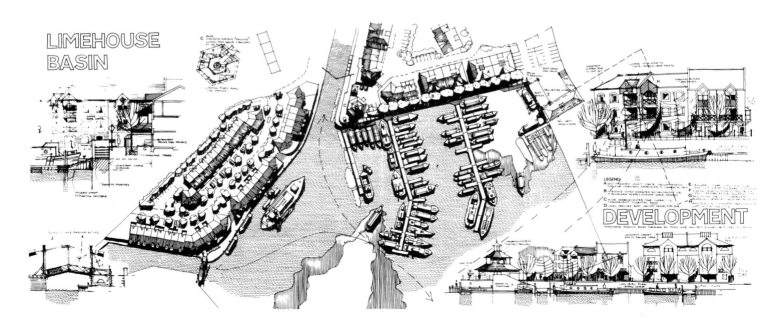

after the canals and rivers under its ownership, the towpaths were not included in the legislation. They were regarded rather as unnecessary appendages – luxuries almost – on which money could be spent but no one in authority would complain if they were neglected. As a result, if British Waterways wants to open up its towpaths to the public – as it does – then it has to spend a good deal of time in complex negotiations with local authorities and other bodies. If new promises are to be kept, then funds have to be found. Government subsidies are not, in the last decade of the twentieth century, the flavour of the age, and the canals are not in themselves big money-earners. But the waterways have at least got a degree of freedom, in that they can keep their own earnings these days, instead of being forced to hand them over to the Treasury, which would then hand them back again (or possibly not). But if traffic – holiday or freight – is scarcely enough to keep the system ticking over, where is the money for improvement to come from? The answer seems to be one that can be given a variety of names: to those who favour the scheme, it is known as capitalizing on one's assets, the realistic use of resources, and other variations on that theme; to critics, it is selling the family silver.

The canal system has developed over a period of more than 200 years, and the system we have today is not the same as the one that was created in the eighteenth century. It is one with new needs and new requirements, and there is a persuasive logic in the view that it has always been subject to change in the past and will be subject to further change in the future. If, to take just one example, the mainte-nance yard at Bull's Bridge on the Grand Union is no longer needed, why should it not be sold off, demolished and replaced by a supermarket? The cash is desperately needed for projects that will earn income for repairs, maintenance and the preser-vation of far more interesting and valuable sites. The yard has, after all, seen more than one use – in an earlier existence it was the Grand Union fleet depot. It is not a site of architectural or historical distinction, could easily be dispensed with, and should be allowed to slip quietly into oblivion. It is also a comfort to know that British Waterways' architects and environmentalists, Waterway Environment Services, vet all such sell-offs, and I, for one, have complete confidence in their integrity and in their concern to preserve the unique nature of our canals. This is not the place to look at individual cases, but it is appropriate to look at the whole pattern of canal development and try and put together an overall view, within which individual decisions can make sense.

It does not seem that long ago that industrial buildings – any industrial buildings – were simply regarded as blots on the landscape. Blake's famous lines about the 'dark Satanic mills' were taken as architectural criticism, rather than as condemnation of the conditions of the age – the spirit in which they were surely intended. I remember

Wigan pier on the Leeds & Liverpool Canal: (left) neglected and unloved, and (right) restored as a museum.

walking through one of the older industrial communities in South Wales, notebook in hand, and being accosted by worried locals demanding to know if I was from the council, and was I planning to knock their homes down? This seemed to reflect a jaundiced, but probably accurate, view of the local council, but also to show an all too prevalent attitude towards buildings of historic interest and importance. Now the pendulum has not so much swung the opposite way, as changed direction. 'Heritage' is the key word, but all too often it does not mean a respect for those survivors of the past that are worthy of preservation, but a wholly artificial glorification of anything historical. It is history without the pain; bearing as much relationship to historical reality as, say, the Victorian image of the Middle Ages, with its handsome knights and distressed maidens. Jolly people in clogs and shawls represent mill workers, with no mention of rickets, lung disease and the horrific saga of industrial accidents. The trouble is that all this has made 'heritage' into something of a dirty word among the intelligentsia, to be spat out along with other disgusting phrases like 'Disney' and 'theme park'. I should like to give a personal view of the canals, their importance and the directions that conservation and development should take. It begins with an assessment of the period in which they were built.

Traditional history books give the dates 1760–1820 for the Industrial Revolution. It is no accident that this coincides with the canal age, for the Industrial Revolution, which saw production move from home to factory, power change from water-wheel and windmill to the steam engine, and which witnessed a steady move from country-

side to town, depended on the availability of an efficient transport system. What is not so immediately obvious is the importance of these changes to world history. It was an extraordinary transformation, the effects of which are still being felt today. And it happened in Britain. Nowadays, we travel the world with ease to gaze at its marvels, but we have only a dim notion of the true significance of the Pyramids of Egypt or the exotic temples of India. That they had immense significance for the people of their day is beyond question; that they can still move us today by their sheer power, expressed in stone, or by the delicacy of their details, is not in doubt. But they were not part of a great movement that was to make a fundamental change to the whole world.

The world that we know is an industrial world. Industry has led to economic growth and if, at the end of the twentieth century, there are those who claim that the price paid for that growth has been too high, there is no argument about whether or not it has occurred. In terms of world history, the remains of Britain's early industries are of absolute prime importance – physical records to put alongside the other records set down in words. It is not always easy for us to see this. We are still too close to events to appreciate their significance. A mill may have been built two centuries ago, but for many it is still just the factory down the road that closed when times got hard. That same mill may have been built alongside what was then the major new transport route of the area, the canal – but the canal is now seen as old, or more likely, 'old-fashioned', and it is difficult to read importance into a stretch of muddy water graced with a film of oil, two supermarket trolleys and a dead dog. But as time goes on, as the world changes yet again, so the importance of the mill and canal will grow. No one would pretend that an old factory will ever be admired in the way that, say, the Taj Mahal is admired, but this account of the canals of the mania years should have helped make clear that they too have their own special appeal. And it is not pitching it too high to say that they have their own aesthetic appeal.

If, as I believe, the canal system was built up as part of what is arguably the most important transformation of human society that has taken place in Britain since the Roman occupation, then that must affect the way we look at it. We can no longer be content with short-term expedients, but must take a longer view. Change must be given this context. This is not at all the same as saying that change should not occur: no one is suggesting the creation of a 2,000-mile-long museum. Still less would one want to see every lock cottage occupied by a costumed actor, who would pop out to provide local colour at regular intervals. What it does mean is that the historical context must be respected. To some extent this already happens. There are over 2,000 structures on the waterways listed as ancient monuments or as historically important. This is excellent, but a building does not exist in isolation: to be fully appreciated, it must have an appropriate context. Blenheim Palace will always be a magnificent building, but it would look very different if it were surrounded by a housing estate instead of parkland. It is easy to list individual parts as worthy of preservation; less easy to keep the integrity of the whole.

It is all very well for armchair critics – including writers – to go on about saving the best of the past while building for the future. But to make sense of the system, it needs to be looked at with a critical, and above all with an intelligent and sympathetic, eye. We need to know what we have got, before we start worrying about what we should keep. In recent years this process has begun. An Architectural Heritage Officer has been appointed by British Waterways – not to sit in an office dreaming up schemes, but to go out and walk the towpaths and to list, examine and assess the various components, from milestones and bollards to warehouses and maintenance yards, that together create the unique canal environment. This is not a job that can be hurried; in fact, it is to run on a five-year timetable. At the same time, others are looking at the natural environment of the waterways and at its ecology. It is all rather like a giant inventory, providing the hard data on which, one devoutly hopes, sensible decisions can be based.

Out of this audit has already come a series of 'corridor studies', carried out by Waterway Environment Services. The aim was to look at long sections of canal, to try and see the overall pattern into which individual developments and changes could be fitted. One study was devoted to the southern Stratford. In 1988 the National Trust finally said goodbye to its canal and handed it over, with (one imagines) a hearty sigh of relief, to British Waterways. As part of the change, the two bodies jointly agreed to allocate three million pounds to improvement and development. Waterway Environment Services came up with views and suggestions, and now it is up to all the various bodies – tourist boards, the waterway authority, local councils and private developers – to work together to help realize the possibilities. The framework, at least, has been set in place. That in itself is a great improvement on earlier times, when there often seemed to be no overall sense of direction or purpose.

A key element in all future development must be the preservation of individuality. One of the joys of canal travel is the movement from one canal to another, and discovering that each canal has its own wholly distinct personality. British Waterways is moving towards a more regionally based form of management. The standards of one canal will not necessarily have to be applied to another. Inevitably, however, there will be changes, and there are three types of change that can easily be identified. The first looks for ways to improve the canals for all their users; the second looks to raise cash by selling off buildings and land that no longer seem to have any use; the third seeks to develop sites in partnership, creating income.

Everyone wants to improve the canals, but what looks like improvement to some can be seen as quite the opposite by others. Take a simple example – safety. Everyone is in favour of safety, but not everyone would agree about the way in which things can be improved. To the casual observer it must look a little startling to see people walking across unguarded lock gates and operating potentially dangerous mechanisms, with no form of safety protection at all. And make no mistake, there are real dangers here, as there are in any activity involving deep water. But the great advantage of the traditional lock is that it does actually look dangerous, so that newcomers approach it with a proper sense of caution. The hydraulic gear that once, but happily no longer, seemed likely to replace all the old-fashioned paddles, looks safer but is not necessarily so. If something does go wrong at a lock, the movement of the water can be stopped virtually on the instant by dropping the old paddles, whereas hydraulics have to be wound slowly down.

Locks are dangerous but essential. There is no possibility of having every lock manned and overseen, and it is difficult to see what can actually be done to make them safer. The present 'everything on show and open' lock does at least look like a threat, and if things should go wrong there is nothing to prevent a quick recovery. I am reminded of a stretch of urban canal, where a local authority in search of safety built a high iron fence between the waterway and the local housing estate. It was well intentioned, but in practice small children could easily squeeze between the railings, and did. Only adults could not get through, or over. A child could reach the water and fall in; an adult could not get there for a rescue.

Again, attitudes are changing. New housing estates, offices and factories seem to want to open out, with waterside lawns and walks. But the old ideas still linger. Citizens of one group of homes on the Paddington arm of the Grand Union have created a Colditz for themselves. First comes the high wooden fence, then the rolls of barbed wire, and finally a chained fence by the towpath, topped with still more barbed wire.

One way in which danger is being minimized is through education. The more that children – and adults – understand and know about the waterway system and how to handle it, the fewer accidents there will be. But no amount of education will ever quite get rid of the cussed determination of kids to do something they know they should not. The best-built boat in the world will never, alas, give as great a thrill as the raft made of two rotting planks inexpertly lashed to a pair of leaking oil drums.

The traditional lock is a vital part of the canal scene. Water has its dangers and they cannot be wished away. But there is another reason for wanting to change the lock scene – to make things easier for holiday boaters. This seems to me to be even less supportable. The canal holiday is essentially an outdoor holiday, and a good part of its pleasure comes from mastering such skills as working efficiently through a lock. Working a lock is really only difficult when it has not been properly maintained. Even if there is a vociferous movement for making life easy, it should be resisted. This is where the broad, historical view comes in again.

The historical outline of the canals of 1793 should have shown that although they were all begun in the one year, they are by no means uniform. Different companies employed different engineers; different engineers had different notions of how things should be done. Canals going through different regions made use of different materials, and so their individual characters were built up. Nowhere is this character seen more clearly than at the lock; abutments may be of brick or stone, textures will vary, the mechanisms will be different. Some locks have double gates at both ends; others have single gates at the top. Bridges can be of wood, iron, brick or stone; some, such as the split bridges of the Stratford, show great ingenuity in their design. But in no single item is change from canal to canal more apparent than in paddle gear. At one extreme is the covered gear of the Grand Union; the rows of black-and-white cylinders leaning away from each other provide instant identification for a flight such as Hatton. Restored canals, which have begged and borrowed paddle gear from wherever it was available, show a bewildering variety of gears, making a journey into a sort of canal quiz – spot the origin. And this is as much a reflection of the canal's history as any other distinctive form of gear. And traditional gear has one great virtue, which its more modern rivals will never match – it makes a most wonderful and satisfactory noise. The chuntering rattle of well-maintained gear echoing across the water is one of the very special sounds of the canal.

A new development of marina and housing on the Grand Union at Milton Keynes.

There are changes to the canals that are still controversial, including the growth of marinas in recent years. Some find them distasteful, unhappy reminders that the old days of the carrying trade on the narrow canals have gone for ever. There is also a certain sense of guilt by association – there are so many marinas along the coast stacked with sleek vessels which, like Gilbert's ruler of the Queen's Navy, 'never go to sea'. They are there to impress, and to provide somewhere to invite friends for drinks. On the canals, however, marinas do fulfil a genuine function. The alternative is to restrict the number of boats that can use the canal, which might be popular with those who are already there but not with those who wish to join them. And if there are no restrictions and no marinas, the inevitable result will be linear moorings, mile after mile of moored craft, the worst answer of them all.

There is a limit to how far one should go in order to make life easy for pleasure-boaters. There is, in any case, a very high proportion of pleasure-boaters who do not necessarily want things made easier, who positively appreciate the chance to enter a world so very different from that in which they earn their living and spend their everyday lives. It should also be remembered that the boating fraternity is only one

A concept study for a marina.

among a group of users, albeit the most important one: the most important because canals were built to carry boats. It is boats that bring them to life; boats that make sense of all the facilities and mechanisms that one finds along the way. There has always been a fascination in seeing boats go through locks. People stand and stare, as they have done since the canals were built. The canal people even coined their own word for them – 'gongoozlers'. It has its place in De Salis's Glossary of Canal Terms, alongside such technical terms as 'fest ropes', 'loodel' and 'stank'.

The best safety measure of all is good maintenance. But maintenance costs money, and that brings in the other part of the equation – property sales and site development. It is a long time since every lock cottage was occupied by a lock-keeper, and for many years they were rented out. This was an obvious and sensible thing to do, since the worst thing that can happen to any property is for it to be left empty – the fabric suffers, the vandals move in. By renting them out, British Waterways kept control: it could ensure that at the very least nothing too hideous was done to the property; at best, it could use its own expertise and resources to ensure that a building was treated in a way that was in keeping with its original design and setting. Sell a property, and that control passes to local planners. There may be – one hopes there are – stringent rules attached to the sale, but making rules is one thing, enforcing them quite another. One can but hope for the best: lock cottages are, quite frequently, modest and lonely places, their canalside location being their main appeal, and people tend to buy them because they like them for what they are. But another, and promising, approach is now being tried. Instead of selling the cottages, why not let them out as holiday homes? The first cottages to be let by British Waterways are in Scotland and Cheshire. Letting ensures control, provides a regular income instead of a one-off payment, and introduces a lot more people to the delights of the canals. It is British Waterways' view that others are as capable as they are of maintaining such properties, particularly when the new owners have the cash and a direct interest in what would otherwise be no more than surplus buildings. There are many examples of well cared-for properties in private ownership, by my own doubts remain. This may well be true in the short term, but I am less convinced that it will still be true decades from now.

Site development is quite a different matter. A supermarket is never going to look like a maintenance yard, no matter how much goodwill there is on the part of the developers. You can see where the attempt has been made at Leighton Buzzard, where a new Tesco stands beside the canal. The company has established its own style of neo-vernacular architecture – single storey, red brick, gables, pantiles, a sort of cross between an overgrown barn on a model farm and a very superior bungalow. The side facing the water has motifs from mill and warehouse, with canopies over non-existent loading bays and imitation hoists, and to complete the effect a narrow-boat weather-vane swings over the roof. The best thing about the scheme is that it does actually acknowledge that the canal exists. There is even a gate to allow boaters to get from the towpath to the store. But already this kind of backward-leaning architecture has become a cliché. Canalside buildings along the Grand Union have taken a bolder approach, and the result has certainly been a good deal more controversial. At Camden Town, in London, a row of gleaming metallic boxes stands out above the water, a group of houses as unlike the traditional notion of a canalside terrace as one can get. They appear halfway between the much admired Little Venice and Islington. It is hard to find anyone who does not enjoy the architecture of the two latter areas, so why put this outlandish chunk of grey, shiny metal in the middle? Well, we are not Georgians – we do not use the materials of two centuries ago, nor their styles. The new terrace is daring, stylish and wholeheartedly belonging to its own age. And it too has the spirit of the working canal. If anyone were building new canalside warehouses, instead of refurbishing old ones, then they might well look much like this. In this busy urban environment, the canal is just one element, but it has become a very important one. The old rail-canal interchange buildings house a market, which really does have something of the excitement of an old-style flea market. The little castellated lock cottage provides a delightful, idiosyncratic note. In fact, the original structure collapsed years ago, and what we see now is not so much a repair job as a complete rebuild. It is a compliment to the skill of those responsible that this 1972 building is listed for preservation. There is room in this area for innovation and surprise.

Anyone who believes that things should be 'left as they were' should remember what the first canal terminal was like at Paddington – little more than a few brick buildings surrounded by trees. One cannot stop change in a city. What the Camden locks area shows is that the canal can make a unique focus of interest, around which all kinds of things can happen. It works largely because the new is still in scale with the old, and partly because for once the imagination of the designers of the new has matched the ingenuity of the creators of the old.

Something of the same sort can be seen happening further up the canal at Uxbridge. There has been an absolute rash of new buildings, including one new block close by the Denham Yacht Station, which seems not to have taken its themes from the modest canal alongside, but to be looking towards something altogether grander. It rears up above the water like the prow of a Cunard liner. There is an air of the 1930s about its curving lines and porthole windows, so that one half expects to hear the band strike up a number by Cole Porter, and see Fred and Ginger waft, light-footed, down the corridors.

It is no bad thing to be jolted occasionally – and the Uxbridge super-liner is a great improvement on the dreary cubes of glass and concrete that make up all too much of post-war development. London and growth areas such as Uxbridge have a dynamism that could never really be contained within the simple scale of most canal building. But there are exceptions and even when development is on a really massive scale, the canal can play its role by bringing things back to a human scale and a human pace. The city that has a canal has a real bonus. Planners often refer to canals in terms such as the 'lungs of the city', which is not a bad image, suggesting space to breathe in freedom. But the same phrase can just as well apply to the municipal park, and the canals offer rather more than that.

One comes back, time and again, to the unique character of a canal – the magic by which a transport system built for work and profit has become so widely admired. It is easy for that character to be swamped by new development, which is why the 'corridor studies' are, potentially at least, so very valuable. But whatever British Waterways may do, it has a direct influence over little more than the waterway itself and its towpath – what happens immediately beyond that is outside its control. Waterways consultative documents have influenced local authorities and affected planning decisions, but the authorities can totally ignore this advice if they choose to do so. There is a very good case for creating a broader corridor, in which planning decisions have to be referred to the waterway authority, as well as to the local council. Many years ago, in his book *Time on the Thames*, Eric de Maré suggested creating a linear national park, based on the River Thames. There is no good reason why something similar should not be adopted for at least some of the canal system. One hesitates to suggest the setting up of any more regulating bodies (the world is too full of committees as it is), but there appears to be a real need for someone to take a broad view of the waterways, and the waterway environment – perhaps a Waterways Conservancy Trust, with sufficient clout to make sure that its views are listened to with care, even when they are not acted upon.

There is an obvious need to protect the canals from wholly inappropriate development on the large scale, but it is just as important to look at the small scale. It is often the minor details that accumulate to give a canal its character, items as small as bridge numbers and mile-posts. Age adds its own shaping to canal furniture: generations of tow ropes have bitten into metal bridge guards to create intricate patterns; bollards have been eaten away to create sculptural forms; steps worn by use have acquired a grace of line that would have surprised their builders. All this goes on without any deliberate intervention. Time does a good job – care is needed only to avoid ruining it. There was a time, when British Waterways first took over the canals, when there was a rush towards uniformity. Everything was painted in the house colours, a truly hideous blue and yellow. Even the narrowboats were affected – out went the old hand lettering of craftsmen, out went roses and castles, and in came dull conformity. That the situation has changed is largely due to the hard work and determination of one of the contributors to this book, Peter White.

Peter White was appointed Chief Architect to British Waterways in 1970, following his successful scheme at Farmer's Bridge on the Birmingham Canal Navigations. He was responsible for what was arguably the best, and certainly the most influential, book on canals during the last 50 years, *The Waterways Environment Handbook*. The handbook was first produced in 1972, but you will not see it on any best seller lists, or in any bookshop. It is a simple volume, a purely practical guide for everyone directly concerned with building, maintenance and engineering along the canals. It covers an astonishing range of subjects. At one end of the scale are amenity symbols to help canal users, telling them where to find everything from the local pub to refuse disposal – pump-in to pump-out. At the other end, there are notes on the best way to develop a city-centre site. But perhaps nothing has had greater effect than the small-scale instructions. Bollard designs can fit their job, blend in with the old and, in turn, look forward to an equally graceful old age. Suggestions are given on canalside planting, on the treatment of verges

British Waterways Board. Waterway Environment Information Sheet.

LOCK GATES/GEAR 03A

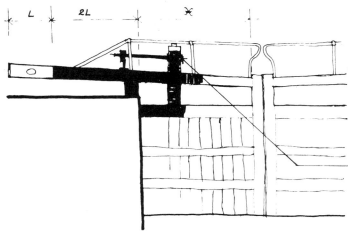

HANDRAILS - White.

BALANCE BEAM Black, with White end. ✗ This part of the balance beam may alternatively be painted White, but other variations are not recommended.

GATE/PADDLE GEARING - Black with White tip.

GATES - Black.

Black stencilled letters
TOP CASING - White

SUPPORT - White.
PADDLE GEAR and this face of support - Black.

ALL - BLACK

BASE - White

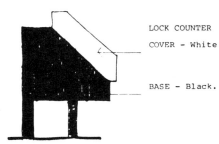

LOCK COUNTER

COVER - White

BASE - Black.

A page from The Waterways Environment Handbook, *showing recommended paint schemes at locks. The* Handbook *is constantly being revised and updated.*

and open areas. Perhaps few decisions have done more to improve the canalside scene than the banishment of blue and yellow and its replacement by an altogether more appropriate colour scheme. Lock cottages have been redecorated in a style that not only suits the individual buildings but reflects something of the character of the canals on which they stand. Iron bridges, such as the famous Horseley Iron Works bridges found on the Oxford and the Birmingham Canals, have appeared in crisp black and white, which emphasizes their form as well as highlighting their details. The list is as long as the book itself, but so often what is said seems no more than common sense – unnecessary, one might almost assume, had experience not shown otherwise. The *Handbook* is not a recipe book for major expenditure, but a guide to how to spend well to the benefit of the canals, how to save money and how to advise others on the best way to spend theirs.

Canal architecture tends to be written about in terms of the very grand structures of the past. Much of the work of Waterway Environment Services is a good deal less glamorous than that. In some cases the greatest compliment that can be paid is to say, 'I didn't notice it.' Canals need a lot of very ordinary facilities – pump-out for toilets, dustbins for rubbish – but no one is likely to admire them. They have to be easy to find by those who need them, yet at the same time unobtrusive, a difficult balancing act. No one gives out awards for a well-built incinerator site – but everyone complains of the unsightliness of a bad one.

Sanitary stations are essential rather than glamorous, but with sensitive treatment they can fit unobtrusively into the canal scene, as here at Brighouse on the Calder & Hebble.

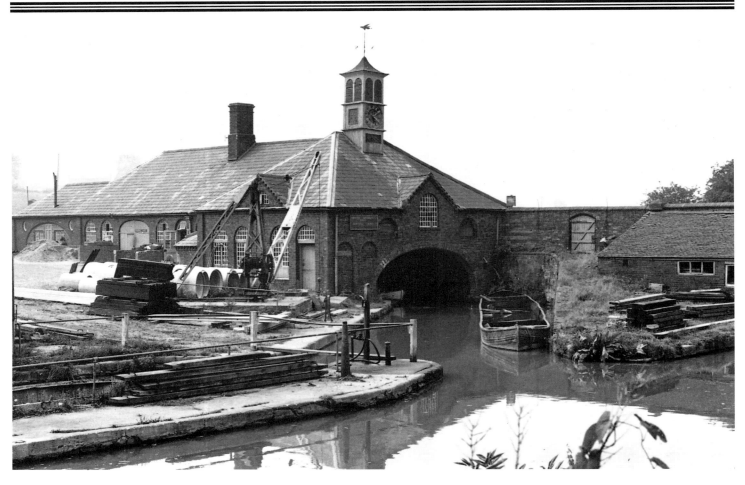

Canal architecture: functional, practical but visually satisfying, Hartshill yard on the Coventry Canal.

Perhaps all this seems to be taking us a long way from the grand dreams of a living monument to the dramatic years of the Industrial Revolution. But one of the lessons that the canal teaches is that, dramatic as it was, its scale was tiny compared with what we are used to today. The rate of change has steadily accelerated, and scale has increased in everything. Six lanes of motorway cross and re-cross the rural windings of Brindley's old canals; more people go over the little Oxford Canal in a day than once travelled on it during a whole year. Even the new generations of canals, born of the mania years of the 1790s, seem no more than romantic anachronisms. It is an effort to wrench the imagination into picturing them as the main commercial arteries of a nation going through the most violent transformation in its entire history. It would be the easiest thing in the world to make that effort impossible in the future. The whole system could be turned into a commercial, linear boat park, with the main criteria the comfort and ease of boaters, and with 'history' reduced to a few colourful set-pieces. Those who are responsible for the system need to have two perspectives: they need to develop a view about the treatment of long stretches of canal, taking the 'corridor' approach; and they need to show an equal concern for the small detail.

There are many people concerned about the future of the canals. Naturalists see them as valuable wildlife habitats. Walkers see them as potential long-distance footpaths, linking country and town. Boaters see them as a place for fun holidays. Anglers see them as elongated fish ponds. All these have a right to their views, and a right to have their opinions taken into account. Yet their views are essentially short-term. The canal system has existed for more than two centuries. In the 1790s men saw it as vital to the whole development of the infant industrial world, and applied all their skills and ingenuity towards making it work. Since then there have been many changes and no doubt there will be many more, but whatever happens nothing can take away the unique historical importance of this system.

We can disguise that importance, mistreat the system, so that nothing of the past can be understood except by reading about it in books. We can attempt to stop time in its tracks, turning the canals into something that has all the life of a stuffed dodo in a museum case. Or we can let the system live by keeping one idea firmly at the front of every other consideration: the canals of Britain are not of passing local interest, not even just of national interest – they are monuments of world importance. We are not their owners – merely their guardians. We can destroy the system, but we cannot re-create it. It is our good fortune that the engineers of 200 years ago built so well. It is our duty now to preserve Britain's canal network for the future.

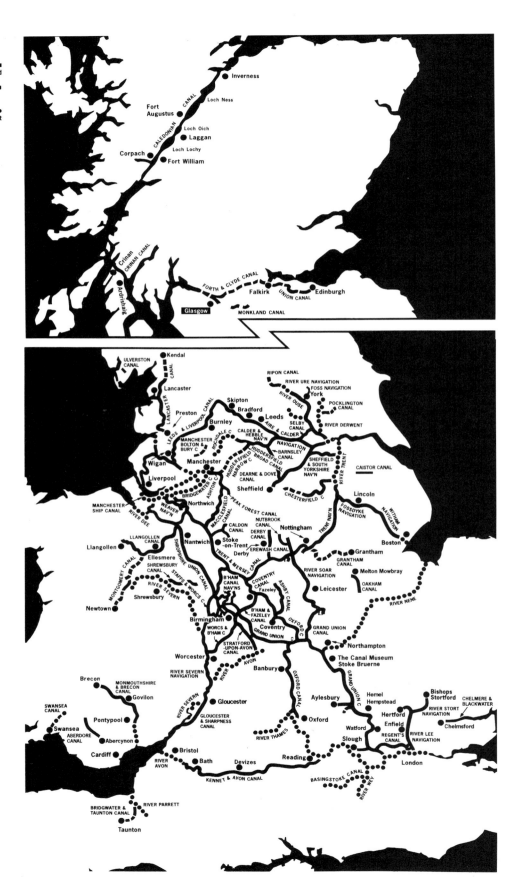

Key

navigable canals and rivers

– – – unnavigable canals and rivers

• • • • canals and rivers not administered by British Waterways

● Towns

Inverness
Loch Ness
CALEDONIAN CANAL
Fort Augustus
Loch Oich
Laggan
Loch Lochy
Corpach
Fort William

Crinan
CRINAN CANAL
Ardrishaig

FORTH & CLYDE CANAL
Falkirk
UNION CANAL
Edinburgh
Glasgow
MONKLAND CANAL

Kendal
ULVERSTON CANAL
LANCASTER CANAL
Lancaster
Preston
Skipton
Bradford
LEEDS & LIVERPOOL CANAL
Burnley
Leeds
ROCHDALE C.
CALDER & HEBBLE NAV'N
AIRE & CALDER
Manchester
MANCHESTER BOLTON & BURY C.
HUDDERSFIELD NARROW C.
HUDDERSFIELD BROAD CANAL
BARNSLEY CANAL
SELBY CANAL
RIPON CANAL
RIVER URE NAVIGATION
FOSS NAVIGATION
York
POCKLINGTON CANAL
RIVER OUSE
RIVER DERWENT
Wigan
ASHTON C.
DEARNE & DOVE CANAL
SHEFFIELD & SOUTH YORKSHIRE NAV'N
RIVER TRENT
CAISTOR CANAL
Liverpool
BRIDGEWATER C.
Northwich
Sheffield
CHESTERFIELD C.
Lincoln
MANCHESTER SHIP CANAL
RIVER DEE
WEAVER NAV'N
PEAK FOREST CANAL
MACCLESFIELD CANAL
CALDON CANAL
NUTBROOK CANAL
DERBY CANAL
Nottingham
TRENT NAV'N
FOSSDYKE NAVIGATION
WITHAM NAVIGATION
LLANGOLLEN CANAL
Nantwich
Stoke-on-Trent
Derby
EREWASH CANAL
Boston
Llangollen
SHREWSBURY CANAL
SHROPSHIRE UNION CANAL
TRENT & MERSEY CANAL
Ellesmere
RIVER SOAR NAVIGATION
Grantham
GRANTHAM CANAL
Melton Mowbray
MONTGOMERY CANAL
STAFFS & WORCS C.
RIVER SEVERN
Shrewsbury
B'HAM CANAL NAV'NS
COVENTRY CANAL
Fazeley
ASHBY CANAL
Leicester
OAKHAM CANAL
Newtown
B'HAM & FAZELEY CANAL
OXFORD C.
RIVER NENE
Birmingham
Coventry
GRAND UNION
GRAND UNION CANAL
Northampton
WORCS & B'HAM C.
STRATFORD-UPON-AVON CANAL
The Canal Museum Stoke Bruerne
Worcester
RIVER AVON
Banbury
Brecon
MONMOUTHSHIRE & BRECON CANAL
Govilon
RIVER SEVERN NAVIGATION
OXFORD CANAL
Bishops Stortford
CHELMERE & BLACKWATER
SWANSEA CANAL
Aylesbury
Hemel Hempstead
RIVER STORT NAVIGATION
Pontypool
Gloucester
GRAND UNION C.
Hertford
ABERDORE CANAL
GLOUCESTER & SHARPNESS CANAL
Oxford
Enfield
Chelmsford
Swansea
Abercynon
RIVER SEVERN
Watford
REGENT'S CANAL
RIVER LEE NAVIGATION
Cardiff
Slough
Bristol
RIVER AVON
Bath
Devizes
RIVER THAMES
Reading
London
KENNET & AVON CANAL
BASINGSTOKE CANAL
RIVER WEY

BRIDGWATER & TAUNTON CANAL
RIVER PARRETT
Taunton

Canals of the Mania

The canal mania that peaked in 1793 was actually spread over three years. The following is a list of the Acts of Parliament passed in those years, but it does not include Acts authorizing minor branches, deviations or those simply empowering the companies to raise extra cash. In one or two cases there were earlier Acts, which were never put into use.

1792

Ashton
Coombe Hill
Horncastle Navigation
Lancaster
Monmouthshire
Nottingham
Wyrley & Essington

1793

Aberdare
Barnsley
Brecon & Abergavenny
Caistor
Chelmer & Blackwater
Crinan
Dearne & Dove
Derby
Dudley No 2
Ellesmere
Foss Navigation
Gloucester & Berkeley
Grand Junction
Grantham
Nutbrook
Old Union
Shrewsbury
Stainforth & Keadby
Stratford-upon-Avon
Ulverston
Warwick & Birmingham

1794

Ashby-de-la-Zouch
Huddersfield
Kennet & Avon
Montgomeryshire
Peak Forest
Rochdale
Sleaford Navigation
Somerset Coal Canal
Swansea
Warwick & Napton
Wisbech

Bibliography

The following is a short list of books that seem particularly relevant to this subject matter.

Burton, Anthony, *The Canal Builders*, David & Charles, 1981 (2nd ed.)
—— *The Great Days of the Canals*, David & Charles, 1989
—— and Pratt, Derek, *Canal*, David & Charles, 1976
Cove-Smith, Chris, *The Grantham Canal Today*, Mitchell, 1974
de Maré, Eric, *The Canals of England*, Architectural Press, 1950
Faulkner, Alan H., *The Grand Junction Canal*, David & Charles, 1972
Hadfield, Charles, *British Canals*, David & Charles, 1984 (7th ed.)
—— and Norris, John, *Waterways to Stratford*, David & Charles, 1968 (2nd ed.)
—— and Skempton, A.W., *William Jessop, Engineer*, David & Charles, 1979
McKnight, Hugh, *The Shell Book of Inland Waterways*, David & Charles, 1981 (2nd ed.)
Paget-Tomlinson, E., *Britain's Canal and River Craft*, Moorland, 1979
Richards, J. M., *The Functional Tradition in Early Industrial Buildings*, Architectural Press, 1958
Rolt, L.T.C., *Navigable Waterways*, Longmans, 1967
—— *Thomas Telford*, Longmans, 1958
Russell, Ronald, *Lost Canals and Waterways of Britain*, David & Charles, 1982
Squires, Roger W., *The New Navvies*, Phillimore, 1983
Stevens, Philip, *The Leicester Line*, David & Charles, 1972
Stevenson, Peter, *The Nutbrook Canal: Derbyshire*, David & Charles, 1970
Tew, David, *The Oakham Canal*, Brewhouse Press, 1968
Ware, Michael E., *Britain's Lost Waterways* (2 vols), Moorland, 1979
White, Peter, *The Waterways Environment Handbook*, British Waterways, 1972 (with regular updates)

Useful Addresses

British Waterways, Willow Grange, Church Road, Watford, Herts, WD1 3QA, tel:
Watford (0923) 226422

British Waterways (Scottish Region), Canal House, Applecross Street, Glasgow, G4
9SP, tel: Glasgow (041) 332 6936

British Waterways (North West Region), Navigation Road, Northwich, Cheshire,
CW8 1BH, tel: Northwich (0606) 74321

British Waterways (North East Region), 1 Dock Street, Leeds, LS1 1HH, tel: Leeds
(0532) 436741

British Waterways (Midlands/South West Region), Peel's Wharf, Watling Street,
Fazeley, Staffs, B78 3QZ, tel: Tamworth (0827) 252000

British Waterways (South East Region), Brindley House, Corner Hall, Lawn Lane,
Hemel Hempstead, Herts, HP3 9YT, tel: Hemel Hempstead (0442) 235400

British Waterways Waterway Environment Services, The Locks, Hillmorton, Near
Rugby, Warwicks, tel: Rugby (0788) 570625

Inland Waterways Association, 114 Regent's Park Road, London, NW1 8UQ, tel:
London (071) 586 2556/586 2510

Waterway Recovery Group, 47 Melfort Drive, Leighton Buzzard, Bedfordshire,
LU7 7XM, tel: Leighton Buzzard (0525) 382311

Index

Picture Credits

The publishers are grateful to the following for permission to reproduce black and white illustrations: Harry Arnold, pp. 218, 219; British Waterways Photo Library, pp. 8, 13, 15, 16, 19, 22, 28, 33, 37, 43, 44, 48, 49, 54, 55, 62, 63, 68, 72, 92, 96, 100, 101, 104, 105, 109, 112, 113, 121, 123, 141, 146, 157, 169, 174, 177, 180, 194, 209, 229, 230; Leslie Bryce, p. 223; Anthony Burton Collection, pp. 52, 53; David McDougall Collection, pp. 41, 81, 106, 129, 132, 136, 152, 161, 166, 181, 188, 189, 202, 205; Ray Merrington Photography, p. 200; National Railway Museum, York, p. 173; National Waterways Museum Archive, pp. 78, 86, 87.